Julia Cookbook

Over 40 recipes to get you up and running with programming using Julia

Jalem Raj Rohit

BIRMINGHAM - MUMBAI

Julia Cookbook

First published: September 2016

Production reference: 1260916

Published by Packt Publishing Ltd.

Livery Place

35 Livery Street

Birmingham B3 2PB, UK.

ISBN 978-1-78588-201-2

www.packtpub.com

Credits

Author

Jalem Raj Rohit

Reviewer

Jakub Glinka

Commissioning Editor

Pratik Shah

Acquisition Editor

Denim Pinto

Content Development Editor

Rohit Singh

Technical Editor

Abhishek R. Kotian

Copy Editor

Pranjali Chury

Project Coordinator

Izzat Contractor

Proofreader

Safis Editing

Indexer

Tejal Daruwale Soni

Production Coordinator

Aparna Bhagat

Cover Work

Aparna Bhagat

About the Author

Jalem Raj Rohit is an IIT Jodhpur graduate with a keen interest in machine learning, data science, data analysis, computational statistics, and natural language processing (NLP). Rohit currently works as a senior data scientist at Zomato, also having worked as the first data scientist at Kayako.

He is part of the Julia project, where he develops data science models and contributes to the codebase. Additionally, Raj is also a Mozilla contributor and volunteer, and he has interned at Scimergent Analytics.

I would thank my parents and my family for all their support and encouragement, which helped me make this book possible.

About the Reviewer

Jakub Glinka is a mathematician, programmer, and data scientist.

He holds a master's degree in applied mathematics from Warsaw University with a specialization in mathematical statistics.

From the beginning of his professional career, he is associated with GfK. His area of expertise ranges from Bayesian modeling to machine learning. He is enthusiastic about new programming languages and currently relying heavily on R and Julia in his professional work.

www.PacktPub.com

For support files and downloads related to your book, please visit www.PacktPub.com.

eBooks, discount offers, and more

Did you know that Packt offers eBook versions of every book published, with PDF and ePub files available? You can upgrade to the eBook version at `www.PacktPub.com` and as a print book customer, you are entitled to a discount on the eBook copy. Get in touch with us at `customercare@packtpub.com` for more details.

At `www.PacktPub.com`, you can also read a collection of free technical articles, sign up for a range of free newsletters and receive exclusive discounts and offers on Packt books and eBooks.

`https://www.packtpub.com/mapt`

Do you need instant solutions to your IT questions? PacktLib is Packt's online digital book library. Here, you can search, access, and read Packt's entire library of books.

Why subscribe?

- Fully searchable across every book published by Packt
- Copy and paste, print, and bookmark content
- On demand and accessible via a web browser

Free access for Packt account holders

Get notified! Find out when new books are published by following `@PacktEnterprise` on Twitter or the Packt Enterprise Facebook page.

Table of Contents

Preface

Julia is a programming language that promises both speed and support for extensive data science applications. Apart from the official documentation of the language, and the individual documentations for each package, there is no single resource that combines all of them and provides a detailed guide to carry out machine learning and data science. So, this book aims to solve the problem by being a comprehensive guide to learning data science for a Julia programmer, right from the exploratory analytics part to the visualization part.

What this book covers

Chapter 1, *Extracting and Handling Data*, deals with the importance of the Julia programming language for data science and its applications. It also serves as a guide to handle data in the most available formats, and shows how to crawl and scrape data from the Internet.

Chapter 2, *Metaprogramming*, covers the concept of metaprogramming, where a language can express its own code as a data structure of itself. For example, Lisp expresses code in the form of Lisp arrays, which are data structures in Lisp itself. Similarly, Julia can express its code as data structures.

Chapter 3, *Statistics with Julia*, teaches you how to perform statistics in Julia, along with the common problems of handling data arrays, distributions, estimation, and sampling techniques.

Chapter 4, *Building Data Science Models*, talks about various data science and statistical models. You will learn to design, customize, and apply them to various data science problems. This chapter will also teach you about model selection and the ways to learn how to build and understand robust statistical models.

Chapter 5, *Working with Visualizations*, teaches you how to visualize and present data, and also to analyze and the findings from the data science approach that you have taken to solve a particular problem. There are various types of visualizations to display your findings, namely the bar plot, the scatter plot, pie chart, and so on. It is very important to choose an appropriate method that can reflect your findings and work in a sensible and an aesthetically pleasing manner.

Chapter 6, *Parallel Computing*, talks about the concepts of parallel computing and handling a lot of data in Julia.

What you need for this book

A beginner level proficiency in the Julia programming language and experience with any programming language, preferably dynamically typed ones such as Python. The software requirements assume you have any of the following OSes: Linux, Windows, or OS X. There are no specific hardware requirements, except that you run and work all your code on a desktop, or a laptop preferably.

Who this book is for

This book is for beginner-level programmers, preferably Julia programmers who are looking to explore and learn the concepts in the domain of data science.

Sections

In this book, you will find several headings that appear frequently (Getting ready, How to do it..., How it works..., There's more..., and See also).

To give clear instructions on how to complete a recipe, we use these sections as follows:

Getting ready

This section tells you what to expect in the recipe, and describes how to set up any software or any preliminary settings required for the recipe.

How to do it...

This section contains the steps required to follow the recipe.

How it works...

This section usually consists of a detailed explanation of what happened in the previous section.

There's more...

This section consists of additional information about the recipe in order to make the reader more knowledgeable about the recipe.

See also

This section provides helpful links to other useful information for the recipe.

Conventions

In this book, you will find a number of text styles that distinguish between different kinds of information. Here are some examples of these styles and an explanation of their meaning. Code words in text, database table names, folder names, filenames, file extensions, pathnames, dummy URLs, user input, and Twitter handles are shown as follows: "The `DataFrames` package is needed to deal with TSV files."

Any command-line input or output is written as follows:

```
query = """ CREATE TABLE Student
        (
            ID INT NOT NULL AUTO_INCREMENT,
            Name VARCHAR(255),
            Attendance FLOAT,
            JoinDate DATE,
            Enrolments INT,
            PRIMARY KEY (ID)
        );"""
```

New terms and important words are shown in bold. Words that you see on the screen, for example, in menus or dialog boxes, appear in the text like this: "If we check the page, we should see our **Page rendered!** alert logged to the console."

Warnings or important notes appear in a box like this.

Tips and tricks appear like this.

Reader feedback

Feedback from our readers is always welcome. Let us know what you think about this book—what you liked or disliked. Reader feedback is important for us as it helps us develop titles that you will really get the most out of.

To send us general feedback, simply e-mail feedback@packtpub.com, and mention the book's title in the subject of your message.

If there is a topic that you have expertise in and you are interested in either writing or contributing to a book, see our author guide at www.packtpub.com/authors.

Customer support

Now that you are the proud owner of a Packt book, we have a number of things to help you to get the most from your purchase.

Errata

Although we have taken every care to ensure the accuracy of our content, mistakes do happen. If you find a mistake in one of our books—maybe a mistake in the text or the code—we would be grateful if you could report this to us. By doing so, you can save other readers from frustration and help us improve subsequent versions of this book. If you find any errata, please report them by visiting http://www.packtpub.com/submit-errata, selecting your book, clicking on the Errata Submission Form link, and entering the details of your errata. Once your errata are verified, your submission will be accepted and the errata will be uploaded to our website or added to any list of existing errata under the Errata section of that title.

To view the previously submitted errata, go to https://www.packtpub.com/books/content/support and enter the name of the book in the search field. The required information will appear under the Errata section.

Piracy

Piracy of copyrighted material on the Internet is an ongoing problem across all media. At Packt, we take the protection of our copyright and licenses very seriously. If you come across any illegal copies of our works in any form on the Internet, please provide us with the location address or website name immediately so that we can pursue a remedy.

Please contact us at copyright@packtpub.com with a link to the suspected pirated material.

We appreciate your help in protecting our authors and our ability to bring you valuable content.

Questions

If you have a problem with any aspect of this book, you can contact us at questions@packtpub.com, and we will do our best to address the problem.

1
Extracting and Handling Data

In this chapter, we will cover the following recipes:

- Why should we use Julia for data science?
- Handling data with CSV files
- Handling data with TSV files
- Working with databases in Julia
- Interacting with the Web

Introduction

This chapter deals with the importance of the Julia programming language for data science and its applications. It also serves as a guide to handling data in the most available formats and also shows how to crawl and scrape data from the Internet.

Data Science pipelines that are used for production purposes need to be robust and highly fault-tolerant, without which the teams would be exposed highly error-prone models. So, these pipelines contain a subprocess called **Extract-Transform-Load** (**ETL**), in which the **Extraction step** involves pulling the data from a source, the **Transform** step involves the transforms performed on the dataset as part of the cleansing process, and the Load step is about loading the now clean data into the local databases for use in production. This will chapter will also teach you how to interact with websites by sending and receiving data through HTTP requests. This would be the first step in any data science and analytics pipeline. So, this chapter will cover some of those methods through which data can be ingested into the pipeline through various data sources.

Why should we use Julia for data science?

Now, you are all set up to learn and experience Julia for data science.

Data Science is simply doing science with data. It applies to a surprisingly wide range of domains, such as engineering, business, marketing, and automotive, owing to the availability of a large amount of data in all these industries from which valuable insights can be extracted and understood.

With the growth of industries, the speed, volume, and variety of the data being produced are drastically increasing. And the tools that have to deal with this data are continuously being adapted, which led to the emergence of more evolved, powerful tools such as Julia.

Julia has been growing steadily as a powerful alternative to the current data science tools. Julia's diverse range of statistical packages along with its powerful compiler features make it a very strong competitor to the current top two programming languages of data science: **R** and **Python**. However, advanced users of R and Python can use Julia alongside each of them to reap the maximum benefits from the features of both.

Julia, with its ability to compile code that looks and reads like Python into machine code that performs like C, has showed a lot of promise with its efficiency at generating efficient code using the type inference. It is also interesting to note that even the core mathematical library of Julia is written in Julia itself. As it supports distributed parallel execution, numerical accuracy, and a powerful type inference, such as Python, and diverse range of statistical packages, such as R, Julia is a very powerful programming language for the very rapidly evolving domain of data science.

Installing and spinning up the Julia terminal is very easy, as follows:

1. Download the `Julia` package suited to your operating system from `http://juli alang.org/downloads/`.
2. Then, fire up Julia's interactive session, which is also called **repl** (**read-eval-print loop**). The terminal output would look like this:

```
               _
           _       _(_)_     |  A fresh approach to technical computing
          (_)     | (_) (_)  |  Documentation: http://docs.julialang.org
           _ _   _| |_  __ _ |  Type "help()" for help.
          | | | | | | |/ _` | |
          | | |_| | | | (_| | |  Version 0.3.2 (2014-10-21 20:18 UTC)
         _/ |\__'_|_|_|\__'_| |  Official http://julialang.org release
        |__/                   |  x86_64-linux-gnu

julia> 2 + 3
5

julia> x = ans
5

julia> y = x * [1 2 3]
1x3 Array{Int64,2}:
 5  10  15

julia> A = [1 2 3; 4 5 6]
2x3 Array{Int64,2}:
 1  2  3
```

3. Installing and spinning up the Julia terminal is very easy:
4. Download the `Julia` package suited to your operating system from
   http://julialang.org/downloads/.

Then, fire up Julia's interactive session, which is also called as repl (read-eval-print loop). The terminal output would look something like this:

Now, you are all set up to learn and experience Julia for Data Science.

Handling data with CSV files

In this section, we will explain ways in which you can handle files with the **Comma-separated Values (CSV)** file format.

Getting ready

Install the `DataFrames` package, which is the Julia package for working with data arrays and dataframes. The command for adding the `DataFrames` packages to the catalog is as follows:

```
Pkg.add("DataFrames")
```

Make sure that all the installed packages are up-to-date: `Pkg.update()`

How to do it...

CSV files, as the name suggests, are files whose contents are separated by commas. CSV files can be accessed and read into the REPL process by executing the following steps:

1. Assign a variable to the local source directory of the file:

   ```
   s = "/Users/username/dir/iris.csv"
   ```

2. The `readtable()` command is used to read the data from the source. The data is read in the form of a Julia DataFrame:

   ```
   iris = readtable(s)
   ```

Data can be written to CSV files from a Julia DataFrame using the following steps:

1. Create a data structure with some data inside it. For example, let's create a two-dimensional dataframe to view the the process of writing files of different formats better using DataFrames:

   ```
   df = DataFrame(A = 1:10, B = 11:20)
   ```

   - The preceding command creates a two-dimensional dataframe with columns named A and B.

2. Now, the dataframe created in Step 1 can be exported to an external CSV file by using the following command:

   ```
   writetable("data.csv", df)
   ```

Handling data with TSV files

In this section, we will explain how to handle **Tab Separated Values** (**TSV**) files.

Getting ready

The DataFrames package is needed to deal with TSV files. So, as it is already installed as instructed in the previous section, we can move ahead and make sure that all the packages are up-to-date with the following command:

```
Pkg.update()
```

How to do it...

TSV files, as the name suggests, are files whose contents are separated by commas. TSV files can be accessed and read into the REPL process by the following method:

1. Assign a variable to the local source directory of the file:

   ```
   s = "/Users/username/dir/data.tsv"
   ```

2. The readtable() command is used to read the data from the source. The data is read in the form of a Julia DataFrame:

   ```
   data = readtable(s)
   ```

Data can be written to TSV files from a Julia DataFrame using the following steps:

1. Create a data structure with some data inside it. For example, let's create a two-dimensional dataframe like the one we created in the previous example:

   ```
   using DataFrames
   df = DataFrame(A = 1:10, B = 11:20)
   ```

2. Now, the dataframe, which we created in Step 1, can be exported to an external TSV file using the following command:

   ```
   writetable("data.csv",df)
   ```

The writetable() command is clever enough to make out the format of the file from the filename extension.

Working with databases in Julia

In this section, we will explain ways to handle data stored in databases: MySQL and PostgreSQL.

Getting ready

MySQL is an open source relational database. To be able to interact with your MySQL databases from Julia, the database server (along with the relevant Julia package) needs to be installed. Assuming that the database is already set up and the MySQL session is already up and running, install the MySQL bindings for Julia by directly cloning the repository:

```
Pkg.clone("https://github.com/JuliaComputing/MySQL.jl")
```

PostgreSQL is an open source object relational database. Similar to the MySQL setup, the server of the PostgreSQL database should be up and running with a session. Now, install the PostgreSQL bindings for Julia by following the given instructions:

1. Install the `DBI` package. The `DBI` package is a database-independent API that complies with almost all database drivers.
2. The `DBI` package from Julia can be installed by directly cloning it from its repository using the following statement:

   ```
   Pkg.clone("https://github.com/JuliaDB/DBI.jl")
   ```

3. Then, install the `PostgreSQL` library by directly cloning the library's repository using the following statement:

   ```
   Pkg.clone("https://github.com/JuliaDB/PostgreSQL.jl")
   ```

4. SQLite is a light, server-less, self-contained, transactional SQL database engine. To interact with data in SQLite databases, one has to first install the SQLite server and make sure that it is up and running and displaying a prompt like this:

```
→ bin sqlite3
SQLite version 3.8.10.2 2015-05-20 18:17:19
Enter ".help" for usage hints.
Connected to a transient in-memory database.
Use ".open FILENAME" to reopen on a persistent database.
sqlite> ▌
```

5. Now, the SQLite bindings for Julia can be installed through the following steps:

   1. Add the SQLite Julia package by running the following standard package installation command:

   ```
   Pkg.add("SQLite")
   ```

How to do it...

Here, you will learn about connecting to databases and executing queries to manipulate and analyze data. You will also learn about the various protocols and libraries in Julia that will help you interact with databases.

MySQL

A MySQL database can be connected by a simple command that takes in the host, username, password, and database name as parameters. Let's take a look at the following steps:

1. First, import the MySQL package:

   ```
   using MySQL
   ```

2. Set up the connection to a MySQL database by including all the required parameters to establish a connection:

   ```
   conn = mysql_connect(host, user_name, password, dbname)
   ```

3. Now, let's write and run a basic table creation query:
   1. Assign the query statement to a variable.

```
query = """ CREATE TABLE Student
            (
                ID INT NOT NULL AUTO_INCREMENT,
                Name VARCHAR(255),
                Attendance FLOAT,
                JoinDate DATE,
                Enrolments INT,
                PRIMARY KEY (ID)
            );"""
```

2. Now to make sure that the query is successfully created, we can get back the response from the connection.

```
response = mysql_query(conn, query)
```

3. Check for a successful connection through conditional statements:

```
if (response == 0)
    println("Connection successful. Table created")
else
    println("Connection failed. Table not created.")
end
```

4. Queries on the database can be executed by the execute_query() command, which takes the connection variable and the query as parameters. A sample SELECT query can be executed through the following steps:

```
query = """SELECT * FROM Student;"""
data = execute_query(conn, query)
```

5. To get the query results in the form of a Julia array, an extra parameter called opformat should be specified:

```
data_array = execute_query(conn, query, opformat = MYSQL_ARRAY)
```

6. Finally, to execute multiple queries at once, use the mysql_execute_multi_query() command:

```
query = """INSERT INTO Student (Name) VALUES ('');
UPDATE Student SET JoinDate = '08-07-15' WHERE LENGTH(Name) > 5;"""
rows = mysql_execute_multi_query(conn, query)
println("Rows updated by the query: $rows")
```

PostgreSQL

Data handling within a PostgreSQL database can be done by connecting to the database. Firstly, make sure that the database server is up and running. Now, the data in the database can be handled through the following procedure:

1. Firstly, import the requisite packages, which are the DBI and the PostgreSQL databases, using the `import` statements:

   ```
   using DBI
   using PostgreSQL
   ```

2. In addition, the required packages for the `PostgreSQL` library are as follows:

   - `DataFrames.jl`: This has already been installed previously.
   - `DataArrays.jl`: This can be installed by running the statement `Pkg.add("DataArrays"))`.

3. Make a connection to a PostgreSQL database of your choice. It is done through the connect function, which takes in the type of database, the username, the password, the port number, and the database name as input parameters. So, the connection can be established using the following statement:

   ```
   conn = connect(Postgres, "localhost", "password", "testdb", 5432)
   ```

4. If the connection is successful, a message similar to this appears on the screen:

   ```
   PostgreSQL.PostgresDatabaseHandle(Ptr{Void}
   @0x00007  fa8a559f160,0x00000000,false)
   ```

5. Now, prepare the query and tag it to the connection we prepared in the previous step. This can be done using the prepare function, which takes the connection and the query as parameters. So, the execution statement looks something like this:

   ```
   query = prepare(conn,  "SELECT 1::int, 2.0::double precision,
   'name'::character varying, " *  "'name'::character(20);"))
   ```

6. As the query is prepared, let's now execute it, just like we did for MySQL. To do this, we have to enter the query variable, which we created in the previous step, into the execute function. It is done as follows:

```
result = execute(query)
```

7. Now that the query execution is over, the connection can be disconnected using the finish and disconnect functions, which take the query and the connection variables as the input parameters, respectively. The statements can be executed as follows:

```
finish(query)
disconnect(conn)
```

8. Now, the results of the query are in the result variable, which can be used for analytics by either moulding it into a dataframe or any other data structure of your choice. The same method can be used for all operations on PostgreSQL databases, which include addition, updating, and deleting.

9. This resource would help you better understand the **Database-Independent API (DBI)**, which we use to connect local PostgreSQL databases such as SQLite.

10. Import the SQLite package into the current session and ensure that the SQLite server is up and running. The package can be imported by running the following command:

```
using SQLite
```

11. Now, a connection to any database can be made through the SQLiteDB() function in Julia Version 3 and the SQLite.DB() function in Julia Version 4.

12. The connection can be made in Julia version 4 as follows:

```
db = SQLite.DB("dbname.sqlite")
```

13. The connection can be made in Julia version 3 as follows:

```
db = SQLiteDB("dbname.sqlite")
```

14. Now, as the connection is made, queries can be executed using the query() function in Version 3 and the SQLite.query() function in Version 4.

- In Version 3:

```
query(db, "A SQL query")
```

- In Version 4:

```
SQLite.query(db, "A SQL query")
```

The `SQLite.jl` package also allows the user to use macros and registers for manipulating and using data. However, the concepts are beyond the scope of this chapter.

So, these are some of the ways through which data can be handled in Julia. There are a lot of databases whose connectors directly connect to DBI, such as SQlite, MySQL, and so on, and through which queries and their execution can be carried out, as shown in the PostgreSQL section. Similarly, data can be scraped from the Internet and used for analytics, which can be achieved through a combination of Julia libraries, but that is beyond the scope of this book.

There's more...

MySQL

The following resource helps you learn more about its advanced features and provides information about the `MySQL.jl` library of Julia. This includes performance benchmarks and details, as well as information on CRUD and testing:

`https://github.com/JuliaDB/MySQL.jl`

PostgreSQL

Visit `https://github.com/JuliaDB/DBI.jl` to understand better the DBI we use to connect local PostgreSQL databases:

Visit `https://github.com/JuliaDB/DBI.jl` for extended and in-depth documentation on the `PostgreSQL.jl` library, which includes dealing with Amazon web services, and so on.

SQLite

Now, as you have learned the ways in which data can be extracted, manipulated, and worked on from various external sources, there are some more interesting things that the database drivers of Julia can do apart from just executing queries. You can find those at `https://github.com/JuliaDB/SQLite.jl/blob/master/OLD_README.md#custom-scalar-functions`.

Interacting with the Web

In this section, you will learn how to interact with the Web through HTTP requests, both for getting data and posting data to the Web. You will learn about sending and getting requests to and from websites and also analyzing those responses.

Getting ready

Start by downloading and installing the `Requests.jl` package of Julia, which is available at `Pkg.add("Requests")`.

Make sure that you have an active Internet connection while reading and using the code in the recipe, as it deals with interacting with live websites on the Web. You can experiment with this recipe on the website `http://httpbin.org`, as it is designed especially for such experiments and tutorials.

This is how you use the `Requests.jl` package and import the required modules:

1. Start by importing the package:

    ```
    Pkg.add("Requests")
    ```

2. Next, import the necessary modules from the package for quick use. The modules that will be used in this recipe are `get`, `post`, `put`, and `delete`. So, this is how to import the modules:

    ```
    import Requests: get, post
    ```

How to do it...

Here, you will learn how to interact with the Web through the HTTP protocol and requests. You will also learn how to send and receive data, and autofill forms on the Internet, through HTTP requests.

GET request

1. The GET request is used to request data from a specified web resource. So, this is how we send the GET request to a website:

   ```
   get("url of the website")
   ```

2. To get requests from a specific web page inside the website, the query parameter of the GET command can be used to specify the web page. This is how you do it:

   ```
   get("url of the website"; query = Dict("title" =>
   "page number/page name"))
   ```

3. Timeouts can also be set for the GET requests. This would be useful for identifying unresponsive websites/web pages. The timeout parameter in the GET request takes a particular numeric value to be set as the timeout threshold; above this, if the server does not return any data, a timeout request will be thrown. This is how you set it:

   ```
   get("url of the website"; timeout = 0.5)
   ```

   - Here, 0.5 means 50 ms.

4. Some websites redirect users to different web pages or sometimes to different websites. So, to avoid getting your request repeatedly redirected, you can set the max_redirects and allow_redirects parameters in the GET request. This is how they can be set:

   ```
   get("url of the website"; max_redirects = 4)
   ```

5. Now, to set the allow_redirects parameter preventing the site from redirecting your GET requests:

   ```
   get("url of the website"; allow_redirects = false)
   ```

   - This would not allow the website to redirect your GET request. If a redirect is triggered, it throws an error.

6. The POST request submits data to a specific web resource. So, this is how to send a post request to a website:

   ```
   post("url of the website")
   ```

7. Data can be sent to a web resource through the POST request by adding it into the data parameter in the POST request statement:

```
post("url of the website"; data = "Data to be sent")
```

8. Data for filling forms on the Web also can be sent through the POST request through the same data parameter, but the data should now be sent in the form of a Julia dictionary data structure:

```
post("url of the website"; data = Dict(First_Name => "abc",
Last_Name => "xyz" ))
```

9. Data such as session cookies can also be sent through the POST request by including the session details inside a Julia Dictionary and including it in the POST request as the cookies parameter:

```
post("url of the website"; cookies = Dict("sessionkey" => "key"))
```

10. Files can also be sent to web resources through the POST requests. This can be done by including the files in the files parameter of the POST request:

```
file = "xyz.jl"
post("url of the website"; files = [FileParam(file), "text/julia",
"file_name", "file_name.jl"])
```

There's more...

There are more HTTP requests with which you can interact with web resources such as the PUT and DELETE requests. All of them can be studied in detail from the documentation for the Requests.jl package, which is available at https://github.com/JuliaWeb/Requests.jl.

2
Metaprogramming

In this chapter, we will cover the following recipes:

- Representation of a Julia program
- Programs for metaprogramming
- Expressions and functions for metaprogramming
- Macros
- Advanced concepts in macros
- Function and code generation

Introduction

Metaprogramming is a concept where by a language can express its own code as a data structure of itself. For example, **Lisp** expresses code in the form of Lisp arrays, which are data structures in Lisp itself. Similarly, even Julia can express its code as data structures.

This makes it possible for Julia to generate and transform code through a Julia program. Julia has really nice reflection properties. So, the property of metaprogramming makes it easy to handle repetitive programming and function execution in data science and, especially, while handling big data in the Map Reduce framework.

Representation of a Julia program

In this section, you will study the life of a Julia program and how it is actually represented and interpreted by Julia. You will also learn what is meant by "a language expressing its own code as a data structure of itself."

This section will act as a foundation for learning about the concept of metaprogramming and how Julia uses it for generating code.

Getting ready

To get started with this section, you must simply have your Julia REPL up-and-running.

How to do it...

Firstly, it is very important to know that every Julia program starts out as a string. Let's consider a short program for adding two variables as our Julia code and use it to learn how Julia interprets programs:

```
code = "a + b"
```

It would look like this:

```
julia> code = "a + b"
"a + b"
```

Now, if you parse the preceding string `code`, it would return an object of type `Expression`. Let's check it by actually parsing an example Julia program and checking for its type:

```
check = parse(code)
```

The output would look like this:

```
julia> check = parse(code)
:(a + b)
```

You will learn more about the preceding notation later. Now, the parsed string check is a Julia expression. You can verify that by checking the type of check using the typeof() function in Julia:

```
typeof(check)
```

The output would look like this:

```
julia> typeof(check)
Expr
```

This will return Expr as an output, which is nothing but a short form of Expression, which is what the current type of check is.

Now you know what an expression is and where it originates, let's take a close look at the expression object. To look at the different parts of an expression, you can use the names() function. It can be done like this:

```
names(Expr)
```

The output would look like this:

```
[julia> names(Expr)
3-element Array{Symbol,1}:
 :head
 :args
 :typ
```

So, the expression object consists of three parts: the head, the arguments, and the type. Each argument can be checked in the following ways:

- The head of the expression can be checked using this:

```
check.head
```

- This returns :call as an output.

```
julia> check.head
:call
```

- The type of the expression can be checked as using this:

  ```
  check.typ
  ```

- This returns `Any` as the output. The type can be either given by the user in the program or inferred by the compiler, as Julia is a dynamic programming language.

```
julia> check.typ
Any
```

- The arguments of the expression can be checked using this:

  ```
  check.args
  ```

- This returns an array of the arguments used in the code. The output looks like this:

```
julia> check.args
3-element Array{Any,1}:
 :+
 :a
 :b

julia>
```

The arguments can be symbols, such as +, or –, literal values, or maybe other expressions.

So, from the preceding process, you can clearly infer that the code in Julia is represented as a data structure of the language itself internally.

Now, let's take a look at the dump () function, which gives a nice, annotated display of the check expression. This can be done by entering the check variable into the dump () function. It can be done using this:

```
dump(check)
```

This should return an output listing the arguments, types, and the head of the expression in an annotated manner. The output would be something similar to this:

```
julia> dump(check)
Expr
  head: Symbol call
  args: Array(Any,(3,))
    1: Symbol +
    2: Symbol a
    3: Symbol b
  typ: Any
```

So, you have seen how Julia interprets code as data structures with a simple example. In fact, you can try the same with a more complicated example like this:

$((a + b)*c) / 6$

This is how to do it:

1. Assign the statement as follows:

   ```
   code = "((a + b) * c) / 6"
   ```

   - This is how it would look along with the output printed underneath:

```
julia> code = "((a + b) * c) / 6"
"((a + b) * c) / 6"
```

2. Parse the program:

   ```
   check = parse(code)
   ```

   - This is how it would look along with the output printed underneath:

```
julia> check = parse(code)
:(((a + b) * c) / 6)
```

3. Check the type of `check`:

    typeof(check)

    - This is how it would look:

```
julia> typeof(check)
Expr
```

4. Now, use the `dump` command to take a look at the contents of the expression:

    dump(check)

    - This is how it would look:

```
julia> dump(check)
Expr
  head: Symbol call
  args: Array(Any,(3,))
    1: Symbol /
    2: Expr
      head: Symbol call
      args: Array(Any,(3,))
        1: Symbol *
        2: Expr
          head: Symbol call
          args: Array(Any,(3,))
          typ: Any
        3: Symbol c
      typ: Any
    3: Int64 6
  typ: Any

julia>
```

How it works...

The entire purpose of metaprogramming is to make Julia express its own code as a data structure of itself. One of the main advantages of this is that complex steps and sophisticated and repetitive tasks can be automated through code generation.

So, to make this possible, every Julia program starts out as a string so that it can be parsed and expressed later.

Then the string that is our program is parsed as an object. And that parsed object is called an **Expression**. Now, the expression object is a store for all the details regarding the parsed string, which is our program. We will later explore how and where the details of the program are stored in the `Expression` objects and how to make use of them.

The conversion or the parsing step, where a piece of code is parsed as an `Expr` object, can be checked by checking their type.

As mentioned, the `Expression` object has complete details about the string, which is our program. This includes the head, type, and argument parts, which would be studied in detail in the next recipe. So, for an annotated view of the details in the `Expression` objects, the `dump` function comes in very handy. It unfurls the details in a nested manner to make understanding complex objects easy.

There's more

For more about about the concept of metaprogramming, visit `https://en.wikipedia.org/wiki/Metaprogramming` and `http://stackoverflow.com/questions/514644/what-exactly-is-metaprogramming`.

For a better understanding of S-expressions, which are used to parse the expressions, visit `https://en.wikipedia.org/wiki/S-expression`.

Symbols and expressions

In this section, you will learn about symbols and expressions in detail. They have a syntactic importance in the metaprogramming concepts of Julia. So, this section would explain them in detail, so as to appreciate the concepts covered so far and those to follow.

Symbols

Symbols are the basic blocks of expressions. They are used for concatenating two strings together. They are also used as interned strings while building expressions.

Getting ready

There aren't any major requirements for this chapter. The only requirement is that your Julia REPL should be up and running.

How to do it...

Symbols can take in some arguments and then return the concatenated string of the string representations of those arguments. This is an example of how you can do it in the REPL:

- symbol("FirstName", "LastName")

    The output of the preceding command would look like this:

```
julia> symbol("FirstName", "LastName")
:FirstNameLastName

julia>
```

- symbol("FirstName", 45)

    The output of the preceding command would look like this:

```
julia> symbol("FirstName", 45)
:FirstName45

julia>
```

- symbol("Foo", :Bar, 86)

    The output of the preceding command would look like:

```
julia> symbol("Foo", :Bar, 86)
:FooBar86

julia>
```

Symbols create interned strings that are used for building expressions. An interned string is an immutable string that is used during string processing for optimizing time and space. The character : is used to create symbols. So, a symbol always starts with a : symbol.

So, this is how you check for a symbol:

```
typeof(:sym)
```

This would return the output `Symbol`.

```
julia> typeof(:sym)
Symbol

julia>
```

How it works...

- Symbols are interned string identifiers. These are the heads of expression objects. Interning a string is a method where one copy of a string is stored. This makes the tasks that need to process the string both time-and space-efficient. Interned strings are also immutable. So, symbols are especially used for performing string processing tasks.
- Symbols can be constructed in two ways. The first way is through the : character, whose significance would be discussed in detail in the next section. Another way to do this is by calling the `symbol()` function.
- The first method is for constructing symbols with a single argument using the : character directly before the argument. The second method is for constructing a symbol from multiple arguments. This can be achieved by enclosing the arguments in the `symbol` function. The resulting symbol would be a concatenated representation of the string representations of the enclosed arguments, as seen in the preceding examples.

There's more

For more about interned strings and the underlying concepts, visit
`https://en.wikipedia.org/wiki/S-expression`.

Quoting

The usage of a semicolon to represent expressions is known as **quoting**. The characters
inside the parentheses after the semicolon constitute an `Expression` object.

How to do it...

To check this behavior, let's check for the type of a similar statement that has an object
inside the parentheses after a semicolon. This can be done in the REPL as follows:

```
typeof(:((a + b) * c) / 6))
```

The preceding command gives the following output:

```
julia> typeof(:( (((a + b)*c))/6) )
Expr
```

Multiple expressions can be represented as a block by quoting them. The syntax would be
as follows:

```
exp = quote
            some code
            some more code
            more code
            a little more
            ...
      end
```

An example with some code inside the code block would look like this:

```
julia> exp = quote
                a = "j"
                b = "u"
                c = "l"
                d = "i"
                e = "a"
            end
quote  # none, line 2:
    a = "j" # none, line 3:
    b = "u" # none, line 4:
    c = "l" # none, line 5:
    d = "i" # none, line 6:
    e = "a"
end
```

Now, let's verify the type of the exp variable with the typeof() function.

```
julia> typeof(exp)
Expr
```

So, the the code block enclosed inside quote and end is indeed an expression.

How it works...

Quoting is the concept of creating expression objects using the : character, as demonstrated in the previous section. They are used to create expressions with a single argument. If the argument is a complex one, it can be enclosed in parentheses and the : character can be used before the parentheses.

As seen in the first recipe, the expression thus constructed from quoting would contain the three features or constituents of an expression. They are the head, arguments, and the type. They can be verified using the dump() function for the expression.

The last type of quoting is through block quotes. This is achieved by including the expressions inside a quote block. As you might have inferred already, this type of quoting is done to build and create `Expression` objects from multiple expressions or lines of code.

Interpolation

Sometimes, construction on `Expression` objects is difficult, especially when you have multiple objects and/or variables. This is used for easy and readable expression construction.

So, interpolation is a way to deal with this. Such objects can be interpolated into the expression construction through a $ prefix. This process is also called **splicing** expressions, variables, or literals into quoted expressions.

How to do it...

Suppose there is a literal p, which has to be interpolated for constructing an expression with other literals; this is how it would be done:

```
p = 6;
exp = :(20 + $p)
```

This is how it would look:

```
[julia> p = 6;

[julia> exp = :(20 + $p)
:(20 + 6)
```

For nested quoting, each symbol must be quoted separately, along with splicing the overall parentheses of the nested expression:

```
p = 6;
q = 7;
:(:p in $( :(:p * :q ) ) )
```

This is how it would look in the REPL:

```
julia> p = 6;

julia> q = 7;

julia> :(:p in $(   :(:p * :q ) ) )
:($(Expr(:in, :(:p), :(:p * :q))))
```

Even data structures can be spliced into an expression construction. Now, let's consider the tuple data structure for splicing into an expression builder:

```
p = 6;
exp = :( p in $ :((1,2)) )
```

The preceding command will give the following output:

```
julia> p = 6;

julia> exp = :( p in $ :((1,2)) )
:($(Expr(:in, :p, :((1,2)))))
```

How it works...

Sometimes, constructing expressions becomes difficult and tedious with quoting and direct construction, especially when dealing with complex expressions, which also include variables. This can be done with interpolation, which can also be called the process of **splicing**. This helps include literals and also other expressions into the expression construction process.

Complex expression construction, like including data structures, such as tuples, into expressions, can be easily done by interpolating. Splicing is done by adding the $ symbol before the expression or the variable, which has to be spliced in the expression construction. This is just like how we performed quoting. Splicing makes the process of quoting easier through easier syntax for complex expression construction.

Splicing comes in very handy especially when including expressions that contain conditionals. In the examples in this recipe, we saw that splicing helps in including a data structure on which a conditional for a variable or a character can be easily applied, which would have been very difficult otherwise with just the basic expression constructors. This also helps in making the process of expression construction easy and readable, which can turn out to be very complex otherwise when using complex data structures in the expression construction process.

There's more

To understand how string interpolation works in Julia, you can explore the following link in Julia docs:

```
http://docs.julialang.org/en/release-.4/manual/strings/#man-string-interpola
tion
```

The Eval function

The `eval()` function is simply used for executing or evaluating an `Expression` object. The evaluations and executions are done in a global scope.

Getting ready

To get started with this section, you must simply have your Julia REPL up and running.

How to do it...

Let's work on some examples to understand the `eval()` function better.

Construct an expression for adding two variables:

1. First define the variable:

    ```
    p = 2
    q = 3
    ```

The output would look like this:

```
[julia> p = 2
2

[julia> q = 3
3
```

2. Now, construct the expression:

```
exp = :(p + q)
```

```
[julia> exp = :(p + q)
:(p + q)
```

3. Now, check the value of the expression with the `eval()` function:

```
eval(exp)
```

```
[julia> eval(exp)
5
```

Now, let's look at functions that take in one or more `Expression` objects as input arguments and return another `Expression` object as the output. Let's understand this better through an example:

1. The following code creates a function that we discussed in the preceding example, one which takes in expressions as inputs and also return expressions as outputs:

```
function example_exp(op, var1, var2)
exp = Expr(:call, op, var1, var2)
return exp
end
```

```
julia> function example_exp(op, var1, var2)
           exp = Expr(:call, op, var1, var2)
           return exp
       end
example_exp (generic function with 1 method)
```

2. Create the necessary expressions by passing on the function created in the preceding example:

```
expression = example_exp(:*, 4, 5)
```

The preceding command will generate the following output:

```
julia> expression = example_exp(:*, 4, 5)
:(4 * 5)
```

3. Now, use the eval() function to evaluate the final result of the expression constructed in the preceding example:

```
eval(expression)
```

```
julia> eval(expression)
20
```

How it works...

The eval() function is used to execute an Expression object. It does so at a global scope. Modules are local variable workspaces in Julia, just like functions in Python. So, all the modules have their own eval() function for evaluating the expressions in a global scope.

The variable inside the expression construction is in a local scope as long as it is not executed through the eval() function. So, the eval() function helps in the robust construction of expressions, be it simple or complex ones. Also, they help to express the results and broadcast them to a global scope.

The `eval` expression can help in generating code for executing in the global scope, which is the basic idea and goal of the concept of metaprogramming. For example, an expression for multiplying two objects can be constructed through any one of the expression construction methods we have already studied. Then, the expression can be used to arbitrarily generate code for any variable when evaluating it. The evaluation would be done by looking up the value of the two variables before the execution is done.

Macros

In this section, you will be introduced to macros, which are used to insert generated code into the programs. So, a macro is simply a block of code that can be compiled directly rather than the conventional method of constructing expression statements and using the `eval()` function. The advantage of using macros is that a block of code that has to be hardcoded multiple times can be generated on-the-fly by creating macros for it.

Getting ready

To get started with this section, you must simply have your Julia REPL up and running.

How to do it...

1. Let's create a macro named `welcome` to print `Welcome to Julia`:

```
macro welcome()
return :(println("Welcome to Julia"))
end
```

This is how it would look when done in the REPL:

```
julia> macro welcome()
           return :(println("Welcome to Julia"))
       end
```

2. Now, let's check the macro you have created in the preceding step. Macros are represented by an @ before their name. So, your macro would be represented by @welcome(). It can be checked as follows:

```
@welcome()
```

This is how it would look when printed in the REPL:

```
julia> @welcome()
Welcome to Julia
```

3. Now, let's include placeholders for variables in the macros. Let's have Julia say Hello <your name>, welcome to Julia:

```
macro welcome(name)
return :(println("Hello ", $name, "Welcome to Julia"))
end
```

This is how it would look in the REPL:

```
julia> macro welcome(name)
          return :(println("Hello ", $name, " Welcome to Julia"))
       end
```

4. Now, let's check by executing the preceding macro in the REPL:

```
@welcome("FooBar")
```

```
julia> @welcome("FooBar")
Hello FooBar Welcome to Julia
```

5. Macro arguments can be viewed using the show() function. This returns all the arguments of a macro function in the form of expressions. This is how to use the show() function:

```
macro lookargs(x)
show(x)
end
```

This is how it would look in the REPL:

```
julia> macro lookargs(x)
           show(x)
       end
```

6. Now, let's use the `lookargs` macro that we created previously:

```
@lookargs(println("I love Julia!"))
```

This is how it would look in Julia:

```
julia> @lookargs(println("I love Julia!"))
:(println("I love Julia!"))
```

How it works...

Macros, as they are generally used in other programs such as Excel, are used for ready-made and quick execution of computations. In this case, a macro is also used similarly. They are used for using generated code for the body of a program. This takes care of the ready-made feature of macros.

They also mask the returned expressions to a tuple of arguments so that they can be compiled directly and to avoid the `eval()` function. This would help in automating repetitive tasks easily and rather quickly, as the hassles of repetitive expression construction can be overcome through one-time macro construction.

The names of macros are denoted by adding the @ symbol before them. So whenever they are used, the macro gets be executed instantly and the result of the execution can be seen.

Macros even allow splicing, which makes them even more convenient for complex macro construction. Using splicing, macros that take complex expressions and variables can be constructed and executed.

Macros are very useful, especially because of the fact that they get executed before the code is parsed. So, they help us conveniently generate code before the code is parsed or before the entire program is run; this saves a lot of time, including the time taken to generate code that has to be typed repetitively otherwise.

Macros can be invoked by calling their name after the @ character, as described previously. The arguments of a macro can also be seen through the `showargs` or the `lookargs` functions. In order to have a detailed view of the macro, one can use the `show()` function, which has the same purpose as the `dump()` function for viewing the details of the expressions.

Metaprogramming with DataFrames

In this section, you will learn about implementing the concept of metaprogramming to dataframes. Dataframes are data structures used for expressing data efficiently. So, using metaprogramming techniques helps speed up the process of dealing with data frames, by automated generation of repetitive tasks and easy syntax.

Getting ready

To get started with this section, you must install the `DataArrays`, `DataFrames`, and `DataFramesMeta` packages of Julia. They can be installed using the `Pkg.add()` function. Check for successful installation by executing the following in the REPL:

```
using DataFrames
using DataArrays
using DataFramesMeta
```

How to do it...

Let's start with the `@with` macro. It is used to express the columns of DataFrames as symbols. Let's verify this and play with the macro. Before that you need to define a DataFrame. Here is how you do it:

```
df = DataFrame(a = [1,2,3], b = [4,5,6])
@with(df, :b + 1)
```

This would add +1 to every value in the *y* column of the data frame df you just created in the preceding example. This would look like this in the REPL:

```
julia> df = DataFrame(a = [1,2,3], b = [4,5,6])
3x2 DataFrames.DataFrame
| Row | a | b |
|-----|---|---|
| 1   | 1 | 4 |
| 2   | 2 | 5 |
| 3   | 3 | 6 |

julia> @with(df, :b + 1)
3-element DataArrays.DataArray{Int64,1}:
 5
 6
 7
```

An important feature in the @with macro is expression wrapping. This would help the compiler understand which expression is to be untouched and which is to be referenced. The expressions wrapped in a ^() function go untouched, as does the Expression object:

```
@with(df, df[:a .> 0, ^(:b)])
@with(df, :a + _I_(:b) )
```

The first statement prints all the elements of b, and the second statement would print the sum of the elements of a and b.

This is how it would look in the REPL:

```
julia> @with(df, df[:a .> 0, ^(:b)])
3-element DataArrays.DataArray{Int64,1}:
 4
 5
 6

julia> @with(df, :a + _I_(:b))
3-element DataArrays.DataArray{Int64,1}:
 5
 7
 9
```

Now, let's take a look at a macro that will help in row and column selection. It helps in selecting the indexes. This is the @ix macro, a name which might be similar if you have a Python background. So, here is how you do it using the dataframe you created in the preceding example:

```
@ix(df, :a .> 1)
@ix(df, :b .> 4)
```

Executing the preceding commands returns the rows after a particular selected row. This is how it looks in the REPL:

```
julia> @ix(df, :a .>  1)
2x2 DataFrames.DataFrame
| Row | a | b |
|-----|---|---|
| 1   | 2 | 5 |
| 2   | 3 | 6 |

julia> @ix(df, :b .>  5)
1x2 DataFrames.DataFrame
| Row | a | b |
|-----|---|---|
| 1   | 3 | 6 |
```

One of the most important steps during a data science experiment is the subsetting of rows in data frames, which are used repeatedly. So, the @where macro helps in automating the this, as follows:

```
@where(df, :a .> 2)
```

This will return all rows that are greater than the given index in the a column in the dataframe df. This is what the output would look like, when executed in the REPL:

```
julia> @where(df, :a .> 2)
1x2 DataFrames.DataFrame
| Row | a | b |
|-----|---|---|
| 1   | 3 | 6 |
```

Column selection and manipulation are very important when data transformations are done on the dataset. So, this can be automated using the @select macro. This is how to use it:

```
@select(df, :a, :b)
@select(df, p = 4 * :a, :b)
```

The first statement would select both the columns in the dataframe. The second statement would multiply both the selected columns by 4. This is what the output would look like, when executed in the REPL:

```
julia> @select(df, :a, :b)
3x2 DataFrames.DataFrame
| Row | a | b |
|-----|---|---|
| 1   | 1 | 4 |
| 2   | 2 | 5 |
| 3   | 3 | 6 |

julia> @select(df, p = 4 * :a, :b)
3x2 DataFrames.DataFrame
| Row | p  | b |
|-----|----|---|
| 1   | 4  | 4 |
| 2   | 8  | 5 |
| 3   | 12 | 6 |
```

There is also a macro for iterating over rows and manipulating them based on control flow statements along with conditional statements. This can be done using the @byrow! macro. This is how it is used:

```
@byrow! df if :a > :b; :b = 2 * :a end
```

This will check whether all the rows of column a are greater than b. And all those which satisfy the condition will be multiplied by 2. This is how it looks when executed in the REPL:

```
julia> @byrow! df if :a > :b; :b = 2* :a end
3x2 DataFrames.DataFrame
| Row | a | b |
|-----|---|---|
| 1   | 1 | 2 |
| 2   | 2 | 4 |
| 3   | 3 | 6 |
```

How it works...

The @with macro helps in dealing with and manipulating the columns of a dataframe. This helps especially during the exploratory data analysis and transformation processes. When a transformation, such as the log transformation or Box-Cox transformation, has to be applied to the column, the @with macro would come very handy, as the code need not be repeated and the required change can be applied immediately.

Expression wrapping features are nice additions to the @with macro, as sometimes we would want to leave a particular value untouched and deal with some values differently. These come in handy especially during processes such as outlier handling and time series trends analytics.

The @ix macro helps in row and column selection. However, it really comes in its own when conditional selection on the dataframe is being done. This is very useful during the exploratory data analytics process. This would also help in slicing and dicing the data according to the required values.

Another very useful macro during the process of exploratory data analytics is the @where macro. It helps in conditional analysis on the data frame; but it also helps in conditional subsetting of the data frame, which can be used in a wide variety of ways in a data science exploratory experiment. It helps in selecting the data which satisfies a particular condition and also subsetting the data frame according to the trends and patterns observed in the dataset.

The @select macro is probably the simplest macro in the DataFramesMeta package. It helps in the selection of one or more columns. It also helps in applying transformations along with the selection of columns. For example, when a data analytics experiment requires a particular column to be multiplied by some value for the purpose of normalization, then this very simple macro would come in very handy for a quick implementation of the task.

The @byrow! macro is used to apply computations to the dataframe row by row. The computations can be complex control flows or transformations. This macro would be very heavily used for data science tasks, such as neural network and error optimization tasks that require constant and multiple iterations. By default, the @byrow! macro is implicitly defined for only a single row; it should be enclosed inside a control flow block.

3
Statistics with Julia

In this chapter, we will cover the following recipes:

- Basic statistics concepts
- Descriptive statistics in Julia
- Deviation metrics
- Sampling
- Correlation analysis

Introduction

In this chapter, you would learn about doing statistics in Julia, along with common problems in handling data arrays, distributions, estimation, and sampling techniques.

Basic statistics concepts

In this recipe, you will learn about the `StatsBase` package, which helps you use basic statistical concepts such as weight vectors, common statistical estimates, distributions, and others.

Getting ready

To get started with this recipe, you have to first install the `StatsBase` package by executing `Pkg.add("StatsBase")` in the REPL.

How to do it...

1. Weight vectors can be constructed as follows:

```
w = WeightVec([4., 5., 6.])
```

```
[julia> w = WeightVec([4., 5., 6.])
StatsBase.WeightVec{Float64,Array{Float64,1}}([4.0,5.0,6.0],15.0)
```

2. Weight vectors also compute the sum of the weights automatically. So, if the sum is already computed, it can be added as a second argument to the vector construction so that it saves time required for computing the sum. Here is how to do it:

```
w = WeightVec([4., 5., 6.], 15.)
```

```
[julia> w = WeightVec([4., 5., 6.], 15.)
StatsBase.WeightVec{Float64,Array{Float64,1}}([4.0,5.0,6.0],15.0)
```

3. Weights can also be simply defined by the weights() function, as follows:

```
w = weights([1., 2., 3.])
```

```
[julia> w = weights([1., 2., 3.])
StatsBase.WeightVec{Float64,Array{Float64,1}}([1.0,2.0,3.0],6.0)
```

4. Some important methods that can be used on weight vectors are:
    1. To check whether the weight vector is empty or not, the isemply() function can be used:

    ```
    isempty(w)
    ```

    2. To check the type of weight values, the eltype() function can be used:

    ```
    eltype(w)
    ```

3. To check the length of the weight vector, the `length()` function can be used:

length(w)

5. There are different kinds of means that can be computed from a dataset, such as the harmonic mean, the geometric mean, and the simple mean. So, the `StatsBase` package has methods to help you do these. This is how to calculate the different means:

1. We can calculate the harmonic mean of *x* with the `harmmean()` function:

harmmean(x)

The output would look like the following:

```
julia> x = [1,2,3]
3-element Array{Int64,1}:
 1
 2
 3

julia> harmmean(x)
1.6363636363636365
```

2. We can calculate the geometric mean of *x* with the `geomean()` function:

geomean(x)

The output would look like the following:

```
julia> x = [1,2,3]
3-element Array{Int64,1}:
 1
 2
 3

julia> geomean(x)
1.8171205928321394
```

3. The general mean can be calculated using the mean() function:

```
mean(x)
```

The output would look like the following:

```
[julia> x = [1,2,3]
3-element Array{Int64,1}:
 1
 2
 3

[julia> mean(x)
2.0
```

4. An extra argument for weights can be added to the mean function to make it the weighted mean:

```
mean(x, weights(w))
```

The output would look like the following:

```
julia> x = [1., 2., 3.]
3-element Array{Float64,1}:
 1.0
 2.0
 3.0

julia> w = WeightVec([4., 5., 6.])
StatsBase.WeightVec{Float64,Array{Float64,1}}([4.0,5.0,6.0],15.0)

julia> mean(x,w)
2.1333333333333333
```

How it works...

The weight vectors are very useful, especially when doing statistical modeling and carrying out machine learning experiments, as they would be used to specify coefficient matrices, the weights for weighted statistical analyses, and so.

So, Julia's Weight vectors help differentiate the actual weight vectors from other normal vectors, which would be very helpful during statistical modeling. Having a pre-calculated value as an argument for the weight vectors is also very helpful, as new variables don't need to be created just to store the sum of the weights, and computational time is saved too.

As weighted means are often used in statistical modeling, having the extra argument in the means function would be very convenient for analysts, as creating new weight vectors and then multiplying the two vectors is often time-consuming and redundant.

The `isempty()`, `eltype()`, and `length()` functions are very important and handy while doing descriptive analytics. They can be used to check the properties and contents of the array quickly.

Means are a common and a very important statistic in the process of exploratory data analytics. They are used to check behavior and also help mine important information from the data.

The harmonic mean is rarely used. It is simply an additive structure, applied on the reciprocals of the observations, and is used in problems that include distance measurement and so on.

The mean, along with other common statistics such as median and mode, gives the analyst a rough idea about the behavior, skewness, and bias of data. It also acts as a preliminary measure and a basis for understanding statistical modeling and machine learning techniques that follow the exploratory analytics process.

Descriptive statistics

Descriptive statistics is the discipline of statistics, where information and features, which explain the essence of data, are extracted and analyzed. This part is very important, as it helps us estimate the shape and features of data for model and algorithm selection.

Getting ready

You have to have the `StatsBase` package ready. This can be done by running `using StatsBase` in the REPL.

How to do it…

1. The variance of a vector can be found using the `var()` function. This can be done by the following:

   ```
   var(x)
   ```

   The output would look like the following:

```
julia> x = [1., 2., 3.]
3-element Array{Float64,1}:
 1.0
 2.0
 3.0

julia> var(x)
1.0
```

2. For calculating the weighted variance of the vector x with respect to weight vector w, both of them can be simply added to the `variance()` function as arguments:

```
julia> w = WeightVec([4., 5., 6.])
StatsBase.WeightVec{Float64,Array{Float64,1}}([4.0,5.0,6.0],15.0)

julia> x = [1., 2., 3.]
3-element Array{Float64,1}:
 1.0
 2.0
 3.0

julia> var(x,w)
0.6488888888888888
```

3. For calculating the standard deviation, the `std()` function can be used. This can be done by executing the following in the REPL:

   ```
   std(x)
   ```

   The output would look like the following:

```
[julia> x = [1., 2., 3.]
3-element Array{Float64,1}:
 1.0
 2.0
 3.0

[julia> std(x)
 1.0
```

4. As with the calculation of the preceding variance, the weighted value of standard deviation with respect to a weight vector *w*, can also be calculated by adding the weight vector as an argument in the `std()` function. This can be done in the REPL:

   ```
   std(x, w)
   ```

   The output would look like the following:

```
[julia> x = [1., 2., 3.]
3-element Array{Float64,1}:
 1.0
 2.0
 3.0

[julia> w = WeightVec([4., 5., 6.])
StatsBase.WeightVec{Float64,Array{Float64,1}}([4.0,5.0,6.0],15.0)

[julia> std(x,w)
0.8055363982396381
```

5. The mean and variance or the mean and standard deviation can also be calculated as a single command. Joint computation would help save time and computation steps, as most statistical analyses need these measures together in most cases.

   This can be done by executing the following statements in the REPL:

   mean_and_std(x,w)

6. The weights argument is optional. So, let's leave it off for these examples:

```
julia> x = [24., 54., 23]
3-element Array{Float64,1}:
 24.0
 54.0
 23.0

julia> mean_and_std(x)
(33.666666666666664,17.61628034896508)
```

   mean_and_var(x,w)

   The output would look like the following:

```
julia> x = [24., 54., 23]
3-element Array{Float64,1}:
 24.0
 54.0
 23.0

julia> mean_and_var(x)
(33.666666666666664,310.3333333333333)
```

7. The skewness of the dataset would be interesting to analyze and would give some really good insights about the way the data is spread and an idea about how to deal with the dataset, keeping the spread information in mind. The skewness of a dataset can be calculated by the skewness () function in the REPL:

   skewness(x)

The output would look like the following:

```
[julia> x = [24., 2., 5, 6., 40.]
5-element Array{Float64,1}:
 24.0
  2.0
  5.0
  6.0
 40.0

[julia> skewness(x)
0.7281201625063924
```

8. Kurtosis measures the similarity of the data set with respect to a normal distribution. The Kurtosis measure of a dataset can be calculated using the kurtosis() function in the REPL in the following way:

```
kurtosis(x)
```

The output would look like the following:

```
[julia> x = [24., 2., 5, 6., 40.]
5-element Array{Float64,1}:
 24.0
  2.0
  5.0
  6.0
 40.0

[julia> kurtosis(x)
-1.098532930672661
```

9. The minimum and maximum values of a dataset can be seen through the extrema() function:

```
extrema(x)
```

The output would look like the following:

```
julia> x = [24., 2., 5, 6., 40.]
5-element Array{Float64,1}:
 24.0
  2.0
  5.0
  6.0
 40.0

julia> extrema(x)
(2.0,40.0)
```

10. The quantiles are the 25%, 50%, 75%, and 100% marks of a sorted vector. So, the quantiles can be calculated using the quantile function:

    quantile(x)

    The output would look like the following:

```
julia> x = [24., 2., 5, 6., 40.]
5-element Array{Float64,1}:
 24.0
  2.0
  5.0
  6.0
 40.0

julia> quantile(x)
5-element Array{Float64,1}:
  2.0
  5.0
  6.0
 24.0
 40.0
```

11. The interquartile range, which means the range of values lying between the third and the second quartiles, can be calculated by the iqr() function:

    iqr(x)

    The output would look like the following:

```
julia> x = [24., 2., 5, 6., 40.]
5-element Array{Float64,1}:
 24.0
  2.0
  5.0
  6.0
 40.0

julia> iqr(x)
19.0
```

12. A summary of all the stats of a vector can be found through the summarystats() function. It returns the values of mean, minimum, maximum, median, first quartile, and third quartile. It would look like this in the REPL:

    summarystats(x)

    The output would look like the following:

```
julia> x = [1., 2., 3.]
3-element Array{Float64,1}:
 1.0
 2.0
 3.0

julia> summarystats(x)
Summary Stats:
Mean:          2.000000
Minimum:       1.000000
1st Quartile:  1.500000
Median:        2.000000
3rd Quartile:  2.500000
Maximum:       3.000000
```

13. This can also be achieved by the `summary()` function:

> **`describe(x)`**

The output would look like the following:

```
julia> x = [1., 2., 3.]
3-element Array{Float64,1}:
 1.0
 2.0
 3.0

julia> describe(x)
Summary Stats:
Mean:          2.000000
Minimum:       1.000000
1st Quartile:  1.500000
Median:        2.000000
3rd Quartile:  2.500000
Maximum:       3.000000
```

How it works...

The mean, variance, and standard deviation of the vectors also take in a weight vector as an optional input argument, just in case you want to calculate the weighted metrics instead of the raw metrics for the input vector.

If you want to have a quick list of mean and variance and/or the standard deviation, then the `mean_and_std()` and `mean_and_var()` functions would be very useful, as they store both the values in a single data structure and can be accessed through simple indexing.

The `describe()` and `summarystats()` functions would be very helpful during initial exploratory analysis, where a quick view of the basic statistics would help you get an idea about the shape and the nature of that dataset.

Variance is a statistic that helps the analyst get an idea about how the data is distributed. Along with the mean and the standard deviation statistics, variance is also one of the most important statistics in the process of exploratory data analytics.

The standard deviation statistic helps in understanding the spread of the data and its observations. This is used in the exploratory analytics step along with the mean and the standard deviation statistics.

The skewness of the data is a highly important measure when carrying out exploratory analytics and the statistical modeling steps. It gives a measure of the spread of the data, and it helps the analyst select an appropriate statistical model for fitting.

The Kurtosis measure is used to learn the shape of the data. It's used like the skewness measure. It tells the analyst how different the shape of the data is when compared to a normal distribution curve.

InterQuartile range measures are similar to the mean statistic, but they also give the 25% and the 75% values, which helps us get a better understanding of the explicit properties of the data set and its skewness.

The `summarystats` and `describe` functions help in getting a detailed summary of the important statistics and measures of the dataset and help the analyst or the data scientist take quick decisions. They aid in the exploratory analytics and statistical modeling steps.

Deviation metrics

Metrics that help calculate the distance or similarity between two vectors are called **deviation metrics**. These metrics help us understand the relationship between the different vectors and the data in them.

Getting ready

You have to have the `StatsBase` package ready. This can be done by running `using StatsBase` in the REPL.

How to do it...

1. For getting the number of elements in a vector that are exactly equal to a set of elements in a vector, we can use the `counteq()` function.

   For example, consider the two vectors: $a = [1, 2, 3, 4, 5, 6]$ and $b = [4, 2, 3, 5, 6, 7]$.

   The elements at the second and third indexes are equal to each other, so they will be returned as a result of the `counteq()` function:

   ```
   counteq(a,b)
   ```

The output would look like the following:

```
julia> a = [1,2,3,4,5,6]
6-element Array{Int64,1}:
 1
 2
 3
 4
 5
 6

julia> b = [4,2,3,5,6,7]
6-element Array{Int64,1}:
 4
 2
 3
 5
 6
 7

julia> counteq(a,b)
2
```

2. The `countne()` function does exactly the opposite of the `counteq()` function in the preceding step. It returns the number of elements that are not equal in both the vectors:

    countne(a,b)

    The output would look like the following:

```
julia> a = [1,2,3,4,5,6]
6-element Array{Int64,1}:
 1
 2
 3
 4
 5
 6

julia> b = [4,2,3,5,6,7]
6-element Array{Int64,1}:
 4
 2
 3
 5
 6
 7

julia> countne(a,b)
4
```

3. The L1 distance is the sum of the absolute differences between the Cartesian co-ordinates. It can be calculated using the `L1dist()` function:

```
L1dist(a, b)
```

The output would look like the following:

```
julia> a = [1,2,3,4]
4-element Array{Int64,1}:
 1
 2
 3
 4

julia> b = [2,3,4,5]
4-element Array{Int64,1}:
 2
 3
 4
 5

julia> L1dist(a, b)
4.0
```

4. Similarly, the L2 distance can also be calculated. It is the square root of the sum of the squared differences between the Cartesian co-ordinates. This can be calculated using the `L2dist()` function:

```
L2dist(a, b)
```

The output would look like the following:

```
julia> a = [1,2,3,4]
4-element Array{Int64,1}:
 1
 2
 3
 4

julia> b = [2,3,4,5]
4-element Array{Int64,1}:
 2
 3
 4
 5

julia> L2dist(a, b)
2.0
```

5. The mean absolute deviation is one of the most commonly used error deviation metrics. It can be calculated using the `meanad()` function:

```
meanad(a, b)
```

The output would look like the following:

```
julia> a = [1,2,3,4]
4-element Array{Int64,1}:
 1
 2
 3
 4

julia> b = [2,3,4,5]
4-element Array{Int64,1}:
 2
 3
 4
 5

julia> meanad(a, b)
1.0
```

6. The mean squared deviation is another widely used metric for calculating error deviation. It can be calculated using the `msd()` function:

```
msd(a, b)
```

The output would look like the following:

```
julia> a = [1,2,3,4]
4-element Array{Int64,1}:
 1
 2
 3
 4

julia> b = [2,3,4,5]
4-element Array{Int64,1}:
 2
 3
 4
 5

julia> msd(a, b)
1.0
```

How it works...

Deviation metrics are very important when estimating errors in the statistics and machine learning models.

The `counteq()` and `countne()` functions help in understanding how different the vectors are, as they give the number of elements that are equal and that are not.

The L1 and L2 distance metrics are very important, especially while calculating error or deviation. They are also widely used in Recommender systems, which find applications in e-commerce platforms such as Amazon, Flipkart, and so on.

The `meanad()` and `msd()` are deviation metrics that find a wide application as accuracy testing metrics for data science models. They are used to test the accuracy between the trained data from the model and the actual test dataset.

Sampling

Sampling is the process where by sample units are selected from a large population for analysis. The sample should be selected such that the results and inferences generated from them should be fairly applicable and can be generalized to the population from which it was initially sampled. There are a lot of ways in which sampling can be done. We will discuss them in the *How to do it...* section.

Getting ready

You have to have the `StatsBase` package ready. This can be done by running `using StatsBase` in the REPL.

How to do it...

1. The simplest way of sampling is random sampling, where one can draw a random element from the array, which is the population. This kind of sampling is generally used for ensuring that selection bias does not happen. This kind of sampling can be done using the `sample()` function of the `StatsBase` package:

```
sample(x)
```

The output would look like the following:

```
julia> x = [1., 2., 3., 4.]
4-element Array{Float64,1}:
 1.0
 2.0
 3.0
 4.0

julia> sample(x)
2.0
```

2. Now, as we have seen how to randomly sample an element from a population, we would now see how to randomly sample *n* elements from the population. This would help us get a sample with multiple elements for studying and mining inferences from it. This sampling technique is also robust to the selection bias. This can be done by adding an argument n to the `sample()` function, where n is the length of the sample being drawn from the population:

   `sample(x, n)`

   The output would look like the following:

```
julia> x = [1., 2., 3., 4.]
4-element Array{Float64,1}:
 1.0
 2.0
 3.0
 4.0

julia> sample(x, 2)
2-element Array{Float64,1}:
 4.0
 4.0
```

3. Now, as we have seen how to sample an array of length *n* from a population of length *x*, a simple tweak to the function would give us the facility to write the sampled elements to a pre-allocated element of length *n*. This can be achieved by the `sample!()` function:

   `sample!(x, n)`

The output would look like the following:

```
julia> a
4-element Array{Float64,1}:
 1.0
 2.0
 3.0
 4.0

julia> x
4-element Array{Float64,1}:
 2.0
 1.0
 2.0
 3.0

julia> StatsBase.sample!(a, x)
4-element Array{Float64,1}:
 4.0
 4.0
 3.0
 4.0
```

4. Direct sampling randomly picks numbers from a given array and stores them in another array. This is simple random sampling, which is the simplest form of sampling. This can be achieved by using the `direct_sample!()` function:

```
direct_sample!(a, x)
```

The output would look like the following:

```
julia> a
4-element Array{Float64,1}:
 1.0
 2.0
 3.0
 4.0

julia> x
4-element Array{Float64,1}:
 4.0
 4.0
 3.0
 4.0

julia> StatsBase.direct_sample!(a, x)
4-element Array{Float64,1}:
 3.0
 4.0
 4.0
 1.0
```

5. Now, let's look at stochastic sampling, which is known as multinomial sampling. Consider two arrays **a** and **x** like in the preceding example. For every element in array **a**, it draws a element from the array, each element having the probability of being drawn as 1/n, n being the total number of elements in the array; it then enter this element into array **x**. The values are inherently ordered. This can be achieved by the xmultinom_sample!() function:

```
xmultinom!(a, x)
```

The output would look like the following:

```
[julia> a
4-element Array{Float64,1}:
 1.0
 2.0
 3.0
 4.0

[julia> x
4-element Array{Float64,1}:
 1.0
 3.0
 4.0
 4.0

[julia> StatsBase.xmultinom_sample!(a, x)
4-element Array{Float64,1}:
 1.0
 2.0
 4.0
 4.0
```

6. For generating a random pair of elements from an array, the samplepair() function can be used:

```
samplepair(a)
```

The output would look like the following:

```
julia> a
4-element Array{Float64,1}:
 1.0
 2.0
 3.0
 4.0

julia> StatsBase.samplepair(a)
(3.0,4.0)
```

7. The next algorithm is the **Knuth's** algorithm. This algorithm is used for random sampling without replacement. As this algorithm/sampling method does not require extra space, it is memory-efficient. It requires only the space equal to the length of the array to be sampled. The Knuth's sampling technique can be achieved through the `knuths_sample!()` function:

   ```
   knuths_sample!(a, x)
   ```

   The output would look like the following:

```
[julia> a
4-element Array{Float64,1}:
 1.0
 2.0
 3.0
 4.0

[julia> x
4-element Array{Float64,1}:
 1.0
 3.0
 4.0
 2.0

[julia> StatsBase.knuths_sample!(a, x)
4-element Array{Float64,1}:
 3.0
 2.0
 1.0
 4.0
```

8. The next sampling algorithm we would be looking at is the **Fisher Yates** algorithm. This algorithm takes in an array of length *n*, and then internally shuffles the indices of the elements. The Fisher Yates algorithm can be achieved in the REPL through the `fisher_yates_sample!()` function:

```
fisher_yates_sample!(a, x)
```

The output would look like the following:

```
julia> a
4-element Array{Float64,1}:
 1.0
 2.0
 3.0
 4.0

julia> x
4-element Array{Float64,1}:
 3.0
 4.0
 2.0
 1.0

julia> StatsBase.fisher_yates_sample!(a, x)
4-element Array{Float64,1}:
 4.0
 3.0
 2.0
 1.0
```

How it works...

Random sampling is one of the simplest and the most used sampling techniques in analytics. It is far less prone to bias, so the analyst does not need to check for sampling bias. So, the `sample()` function in Julia gives us a quick way to sample an element from an array randomly.

The `random()` function used for carrying out random sampling, as we saw in the preceding point, can also be extended to sample a set of elements from the array. An extra argument in the function *n* can be added, which can tell the length or the number of elements we want to sample out from the array. This feature is very useful especially when the analyst wants a randomly sampled subsets of data, instead of a single element as a sample.

Now, the `sample!()` function does the same operation as the `sample()` function we saw earlier, but it stores the sampled elements in a pre-defined Julia array. It just frees the analyst from writing code to initialize arrays for storing the sampled set of elements.

If we want to store the sampled array as a new array, we can use the `direct_sample!()` function. This is similar to the random sampling algorithm earlier. Randomly sampled elements are added to a specified array.

The xmultinomial sampling algorithm is where each element is sampled from a separate binomial distribution. The elements are ordered prior to sampling to optimize computational time and effort.

The `samplepair()` function samples two elements randomly and stores them as a pair. This is especially useful when a pair of random elements is needed for initializing weights for a simple neural network or weighting a statistical model ensemble.

The next algorithm we would look Knuth's algorithm **S** for sampling without replacement. The algorithm works as follows:

1. Select the first n elements from the array.
2. Where $i>n$, for every i^{th} element, have a random chance of n/i of keeping it. Failing this, the sample remains the same. If not, have it randomly ($1/n$) replace one of the previously selected n elements of the sample.
3. Now, run Step 2 for the remaining elements in the array.
4. The last sampling algorithm we would look, is the Fisher-Yates algorithm. The algorithm works as follows:

- **8.1** : Firstly, create an array with the required length, or the required number of indices.
- **8.2**:

```
for i = 1:k
  swap inds[i] with a random one in inds[i:n]
  set x[i] = a[inds[i]]
end
```

Correlation analysis

Correlation analysis is the process that indicates the similarity and relationship between two random variables. For time series data, correlation analysis would be done between two sets of the datasets. And in non time-series data, correlation analysis would generally be done between two independent variables in the dataset. So, in this recipe, we would look at the correlation analysis of time series (signals).

Getting ready

You have to have the `StatsBase` package ready. This can be done by running:

```
using StatsBase
```

How to do it...

1. Autocovariance is the covariance of a piece of time data with itself at two different time points. This helps in understanding how correlated the time series is, with respect to the time dimension. This metric helps us find trends across the time dimension of a time series across different time points. It can be calculated with default lags using the `autocov()` function:

   ```
   autocov(x)
   ```

   The output would look like the following:

```
julia> x = [1., 2., 3., 4.]
4-element Array{Float64,1}:
 1.0
 2.0
 3.0
 4.0

julia> autocov(x)
4-element Array{Float64,1}:
  1.25
  0.3125
 -0.375
 -0.5625
```

2. Now, if we want to set user-specified lags in our autocovariance calculations, we can use an extra argument. This feature in the function would make the calculations and modeling very flexible and convenient for the analyst. This can be achieved by adding the lags in an array, as an argument in the autocovariance function in the following way:

```
autocov(x, [lags])
```

The output would look like the following:

```
julia> x
4-element Array{Float64,1}:
 1.0
 2.0
 3.0
 4.0

julia> autocov(x, [2, 3])
2-element Array{Float64,1}:
 -0.375
 -0.5625
```

3. Similar to the `autocor()` function, we can also calculate the autocorrelation of the time series with itself at different points across the time dimension. This would help in finding highly correlated parts of the series and help the analyst in finding patterns more efficiently:

```
autocor(x)
```

The output would look like the following:

```
julia> x
4-element Array{Float64,1}:
 1.0
 2.0
 3.0
 4.0

julia> autocor(x)
4-element Array{Float64,1}:
  1.0
  0.25
 -0.3
 -0.45
```

4. Custom-defined lags can also be added as an argument to the `autocor()` function, just like we did in the `autocor()` function. They need to be passed in as an array, in the following way:

```
autocor(x, [lags])
```

The output would look like the following:

```
julia> x
4-element Array{Float64,1}:
 1.0
 2.0
 3.0
 4.0

julia> autocor(x, [2, 3])
2-element Array{Float64,1}:
 -0.3
 -0.45
```

5. Cross Covariance is the covariance between two time series datasets at specific time lags. This is used for comparing two time series with each other. This is done by calculating the covariance between the different columns of the two datasets, using the `crosscov()` function in the following way:

    crosscov(x, y)

    The output would look like the following:

```
julia> x
4-element Array{Float64,1}:
 1.0
 2.0
 3.0
 4.0

julia> y
4-element Array{Float64,1}:
 4.0
 7.0
 3.0
 2.0

julia> crosscov(x, y)
7-element Array{Float64,1}:
  0.0
  1.125
  0.0
 -1.25
 -1.25
  0.625
  0.75
```

6. Just like we have seen with the preceding `autocovariance()` functions, we can also add specific lags for the calculating `crosscov()` function by passing the array of lags as an argument in the `crosscov()` function, in the following way:

    crosscov(x, y, [lags])

The output would look like the following:

```
julia> x
4-element Array{Float64,1}:
 1.0
 2.0
 3.0
 4.0

julia> y
4-element Array{Float64,1}:
 4.0
 7.0
 3.0
 2.0

julia> crosscov(x, y, [2, 3])
2-element Array{Float64,1}:
 0.625
 0.75
```

7. Similar to the cross covariance metric, we can also calculate the correlation between two time series data across multiple data points across the time dimension. This can be done with the crosscor() function in the following way:

   crosscor(x, y)

   The output would look like the following:

```
julia> x
4-element Array{Float64,1}:
 1.0
 2.0
 3.0
 4.0

julia> y
4-element Array{Float64,1}:
 4.0
 7.0
 3.0
 2.0

julia> crosscor(x, y)
7-element Array{Float64,1}:
  0.0
  0.537853
  0.0
 -0.597614
 -0.597614
  0.298807
  0.358569
```

8. The autocorrelation metric can be calculated with specific lags, which can be added as an argument in the crosscor() function we have seen previously, in the form of an array:

   crosscor(x, y, [lags])

The output would look like the following:

```
julia> x
4-element Array{Float64,1}:
 1.0
 2.0
 3.0
 4.0

julia> y
4-element Array{Float64,1}:
 4.0
 7.0
 3.0
 2.0

julia> crosscor(x, y, [2, 3])
2-element Array{Float64,1}:
 0.298807
 0.358569
```

How it works...

The autocovariance metric helps us identify the pieces of data in the time series, which have a high level of covariance between them. A high level of covariance is not really a good sign when the analyst is dealing with independent variables and feature selection. It would result in redundant data and information being skewed while developing the predictive model. The formula for autocovariance is as follows:

$$C_{XX}(t,s) = \text{cov}(X_t, X_s) = E\big[(X_t - \mu_t)(X_s - \mu_s)\big] = E[X_t X_s] - \mu_t \mu_s$$

Where **t** and **s** are different points in the time dimension.

In the preceding formula, the lags have been automatically set. However, as we saw in the *How to do it...* section, we can manually set the number of lags where we want to calculate the autocovariance metric. This means that we are doing the same calculations, but now with lag values manually set by the analyst. This would greatly improve the flexibility and extensibility of the function, where the analyst can easily test the covariance for a lag value of their choice.

The autocorrelation metric helps test out the parts where the signal is correlated with itself. As explained in the precious section, sometimes correlated data would result in redundancy, which affects the accuracy of the statistical model. It helps in identifying periodicity and repetitive patterns.

The formula for autocorrelation is as follows:

$$R_{ff}(\tau) = \left(f * g_{-1}\left(\overline{f}\right)\right)(\tau) = \int_{-\infty}^{\infty} f(u+\tau)\,\overline{f}(u)\,du = \int_{-\infty}^{\infty} f(u)\,\overline{f}(u-\tau)\,du$$

Where **t** and **tau** are different points in the time dimension.

Now, if we want to find the correlation between two parts of a signal, the `autocor()` function allows us to do that by letting us add in an extra argument in the form of an array, where the lags we want to analyze at are mentioned. This would make the work of the analyst easy and convenient, and the function highly flexible.

The Cross Covariance metric is similar to the autocovariance metric, but it is calculated across two signals instead of on parts of a single one. The properties of autocovariance apply to this metric also. However, this helps in detecting signals having high covariance with respect to each other. The formula for cross covariance is as follows:

$$C_{XY}(t,s) = \text{cov}\left(X_t, Y_s\right) = E\left[\left(X_t - \mu_t\right)\left(Y_s - v_s\right)\right] = E\left[X_t Y_s\right] - \mu_t v_s$$

Where **t** and **s** are different points on different time series/signals in the time dimension.

Now, similar to the other metric we saw previously, we can add custom lags to the cross covariance function as an argument in the form of an array. Doing so would bring in the same advantages as discussed in the preceding points.

The Cross Correlation metric is similar to the autocorrelation metric, but it is calculated across two signals instead of on parts of a single one. The properties of autocorrelation apply to this metric also. This helps in detecting the periodicity, similarity, and trends of a signal or time series with respect to the other. The formula for cross correlation is as follows:

$$(f * g)(\tau) \overset{\text{def}}{=} \int_{-\infty}^{\infty} f^*(t)\,g(t+\tau)\,dt$$

Where **f** and **g** are two signals or time series.

Now, similar to the other metric we saw previously, we can add custom lags to the `crosscor()` function as an argument in the form of an array. Doing so would confer the same advantages as discussed in the preceding points.

4
Building Data Science Models

In this chapter, we will cover the following recipes:

- Dimensionality reduction
- Linear discriminant analysis
- Data preprocessing
- Linear regression
- Score-based classification
- Clustering
- Bayesian basics
- Time series analysis

Introduction

In this chapter, you will learn about various data science and statistical models. You will learn to design, customize, and apply them to various data science problems. This chapter will also teach you about model selection and the ways to build and understand robust statistical models.

Dimensionality reduction

In this recipe, you will learn about the concept of **dimensionality reduction**. This is the set of algorithms used by statisticians and data scientists when data has a large number of dimensions. It helps make computations and model designing easy. We will use the **Principal Component Analysis (PCA)** algorithm for this recipe.

Getting ready

To get started with this recipe, you have to have the MultivariateStats Julia package installed and running. This can be done by entering Pkg.add("MultivariateStats") in the Julia REPL. When using it for the first time, it might show a long list of warnings; however you can safely ignore them for the time being. They in no way affect the algorithms and techniques that we will use in this chapter.

How to do it...

1. Firstly, let's simulate about a hundred random observations, as a training set for the PCA algorithm which we will use. This can be done using the randn() function:

```
X = randn(100,3) * [0.8 0.7; 0.9 0.5; 0.2 0.6]
```

```
[julia> X = randn(100,3) * [0.8 0.7; 0.9 0.5; 0.2 0.6]
100x2 Array{Float64,2}:
 -0.214421    -0.651294
  0.192329     0.310959
  0.875353     0.439359
 -0.276685    -0.016095
 -0.0384189    0.42374
  3.49147      2.35909
 -2.12082     -2.06463
  1.25676      0.965947
  0.441123     0.311001
 -0.587929    -0.840313
  ⋮
 -1.18698     -1.70781
  0.858047     0.627955
  2.43481      1.48533
  0.897603     1.40961
 -1.43061     -1.61424
 -0.54459      0.223121
  0.295887     0.360804
  1.66216      0.863761
  0.657583     0.1513
```

2. Now, to fit the PCA algorithm on the simulated dataset, we can use the `fit()` function. This fits the PCA algorithm with a given dimension on the dataset with the custom parameters of the user. This can be done as follows:

```
P = fit(PCA, X, maxoutdim = 20)
```

Here, the `maxoutdim` parameter is just the maximum dimension of the output dataset.

```
julia> P = fit(PCA, X, maxoutdim = 20)
PCA(indim = 100, outdim = 1, principalratio = 1.00000)
```

3. Now, as we have created an instance of the PCA algorithm fitted on our dataset, we can explore its properties to understand the algorithm and its Julia implementation better. Firstly, to look at the dimension of the input dataset, which can also be called the **observation space**, we can use the `indim()` function, as follows:

```
indim(P)
```

```
julia> indim(P)
100
```

4. Now, we already know the dimension of the input data. So, we have to look at the dimension of the output data to check whether the algorithm has reduced the dimension of our dataset. This can be done using the `outdim()` function, as follows:

```
outdim(P)
```

```
julia> outdim(P)
1
```

5. Similarly, the mean vector of the input data can be calculated by simply using the `mean()` function on the PCA instance that we have already created. This can be done as follows:

mean(P)

```
julia> mean(P)
100-element Array{Float64,1}:
  0.19502
 -0.620652
  0.766694
  0.852925
  1.47081
  0.249772
 -0.0837288
  1.12172
 -0.492007
  ⋮
  1.40367
 -0.831595
 -1.04743
  0.980702
 -0.909524
  0.529424
 -0.728779
  0.490002
  1.70647
```

6. Now to get the projection matrix that corresponds to the principal components returned by our PCA instance. In the case of multiple principal components, they are arranged in the descending order of their respective variance values.
The `projection()` function helps us get the projection matrix of the instance, as follows:

projection(P)

```
julia> projection(P)
100x1 Array{Float64,2}:
 -0.150278
  0.0370419
  0.156628
  0.0525754
 -0.0527251
  0.049752
  0.0202521
 -0.0837644
 -0.0959538
 -0.0778109
  ⋮
 -0.104665
  0.0557597
 -0.104894
 -0.0286959
  0.0603532
  0.0514137
  0.00725397
 -0.106828
  0.0295974
```

7. To know the variances of the principal components, the `principalvars()` function can be used. We will get a list of the variances of all the principal components of the output data, as follows:

```
[julia> principalvars(P)
1-element Array{Float64,1}:
 8.59967
```

8. To know the total variance of the principal components of the instance, the `tprincipalvar()` method can be used. This will give us an idea of the amount of variance being captured by the selected principal components in the output data, as follows:

```
    tprincipalvar(P)
```

```
[julia> tprincipalvar(P)
8.599668429748876
```

How it works...

The `randn()` function simulates a set of random numbers, which we will use later in the PCA exercise. Here, we generate about a hundred random observations, which we will fit into the PCA algorithm for the dimensionality reduction task.

The `fit()` function helps fit the PCA algorithm to the input dataset, which we simulated in the preceding step. The desired dimension of the output data can also be included as a parameter to the `fit()` function. This will help in deciding the number of principal variance vectors.

The `indim()` function gives the dimension of the input data. It is the raw data to which we want to apply our dimensionality reduction algorithm. In the preceding example, we took a dataset of 100 dimensions, which can be verified by the `indim()` function as shown in Step 3.

The `outdim()` function gives the dimension of the output dataset. It is the dimension of the dataset after the PCA algorithm is fitted on the input dataset, that is, after the dimensionality reduction process is done. In the preceding example, the dimension of the output dataset is 1, which can be verified by the `outdim()` function as shown in Step 4.

The mean of the input data is the like the center of gravity of the data. It is the point or vector, depending on the dimension of the data, around which the entire dataset is spread.

The projection is the scaled or reduced version of the input data. It contains the shape and dimensions of the output data as specified.

The principal variances are the variance values of the principal components of the input data, or the variances of the dimensions of the output data, which is the projection which we saw in the preceding example.

Sometimes, we are interested in knowing the total variance being exhibited by the principal components of the data. This will help the analyst gauge the properties of the data and act accordingly. This will simply be the sum of the variances of the principal components, which we have seen in the previous step.

Linear discriminant analysis

Linear discriminant analysis is the algorithm that is used for classification tasks. This is often used to find the linear combination of the input features in the data, which can separate the observations into classes. In this case, it would be two classes; however, multi-class classification can also be done through the discriminant analysis algorithm, which is also called the **multi-class linear discriminant analysis algorithm**.

Getting ready

To get started with this recipe, you have to clone the `DiscriminantAnalysis.jl` library from GitHub. This can be done by the following command:

```
Pkg.clone("https://github.com/trthatcher/DiscriminantAnalysis.jl.git")
```

And then, we can import the library by calling by its name, which is `DiscriminantAnalysis`. This can be done as follows:

```
using DiscriminantAnalysis
```

We also have to use the `DataFrames` library from Julia. If this library doesn't exist in your local system, it can be added by the following command:

```
Pkg.add("DataFrames")
```

How to do it...

1. Firstly, let's download the famous `iris` dataset. This is a dataset that contains the details of iris flowers and the classes they belong to. This dataset can be used as a nice, simple start to classification problems. It can be downloaded from the following link:

   ```
   https://raw.githubusercontent.com/trthatcher/DiscriminantAnalysis.
   jl/master/example/iris.csv
   ```

   Now, we would like to import the dataset using the `readtable()` function of the `DataFrames` library. This can be done in the REPL, as follows:

   ```
   df = readtable("iris.csv")
   ```

   ```
   julia> df = readtable("iris.csv")
   150x5 DataFrames.DataFrame
   | Row | SepalLength | SepalWidth | PetalLength | PetalWidth | Species     |
   |-----|-------------|------------|-------------|------------|-------------|
   | 1   | 5.1         | 3.5        | 1.4         | 0.2        | "Setosa"    |
   | 2   | 4.9         | 3.0        | 1.4         | 0.2        | "Setosa"    |
   | 3   | 4.7         | 3.2        | 1.3         | 0.2        | "Setosa"    |
   | 4   | 4.6         | 3.1        | 1.5         | 0.2        | "Setosa"    |
   | 5   | 5.0         | 3.6        | 1.4         | 0.2        | "Setosa"    |
   | 6   | 5.4         | 3.9        | 1.7         | 0.4        | "Setosa"    |
   | 7   | 4.6         | 3.4        | 1.4         | 0.3        | "Setosa"    |
   | 8   | 5.0         | 3.4        | 1.5         | 0.2        | "Setosa"    |
   ⋮
   | 142 | 6.9         | 3.1        | 5.1         | 2.3        | "Virginica" |
   | 143 | 5.8         | 2.7        | 5.1         | 1.9        | "Virginica" |
   | 144 | 6.8         | 3.2        | 5.9         | 2.3        | "Virginica" |
   | 145 | 6.7         | 3.3        | 5.7         | 2.5        | "Virginica" |
   | 146 | 6.7         | 3.0        | 5.2         | 2.3        | "Virginica" |
   | 147 | 6.3         | 2.5        | 5.0         | 1.9        | "Virginica" |
   | 148 | 6.5         | 3.0        | 5.2         | 2.0        | "Virginica" |
   | 149 | 6.2         | 3.4        | 5.4         | 2.3        | "Virginica" |
   | 150 | 5.9         | 3.0        | 5.1         | 1.8        | "Virginica" |
   ```

2. Next, we pool the dataframe's **Species** column, which contains the names of the types of flower. Pooling a vector converts the string values into factors. This can be done using the `pool!()` function. The `!` is used as the function changes the value of its arguments after execution. So, this can be done as follows:

```
pool!(df, [:Species])
```

```
julia> pool!(df, [:Species])
```

3. Now, convert the array of the independent variables into the type `Float64` for generality. This will help us avoid problems related to the *type* later in the analysis. It can be done using the `convert()` function, which takes in the dataframe columns and the type, which it has to be converted to. This can be done as follows:

```
X = convert(Array{Float64}, df[[:PetalLength, :PetalWidth,
:SepalLength,
        :SepalWidth]])
```

```
julia> X = convert(Array{Float64}, df[[:PetalLength, :PetalWidth, :SepalLength,
:SepalWidth]])
150x4 Array{Float64,2}:
 1.4  0.2  5.1  3.5
 1.4  0.2  4.9  3.0
 1.3  0.2  4.7  3.2
 1.5  0.2  4.6  3.1
 1.4  0.2  5.0  3.6
 1.7  0.4  5.4  3.9
 1.4  0.3  4.6  3.4
 1.5  0.2  5.0  3.4
 1.4  0.2  4.4  2.9
 1.5  0.1  4.9  3.1
 ⋮
 5.1  2.3  6.9  3.1
 5.1  1.9  5.8  2.7
 5.9  2.3  6.8  3.2
 5.7  2.5  6.7  3.3
 5.2  2.3  6.7  3.0
 5.0  1.9  6.3  2.5
 5.2  2.0  6.5  3.0
 5.4  2.3  6.2  3.4
```

4. Now, let's define the dependent variable, which is the **Species** labels of the dataset. We assign it to a variable y, as per the general practice. This can be done as follows:

```
y = df[:Species].refs
```

```
julia> y = df[:Species].refs
150-element Array{UInt8,1}:
 0x01
 0x01
 0x01
 0x01
 0x01
 0x01
 0x01
 0x01
 0x01
 0x01
    ⋮
 0x03
 0x03
 0x03
 0x03
 0x03
 0x03
 0x03
 0x03
 0x03
```

5. Now, we can fit our linear discriminant classification model onto the dataset we have constructed in the preceding processes. It can be done using the lda() function with the parameters being the datasets of the dependent and the independent variables, as follows:

```
model = lda(X, y)
```

```
julia> model = lda(X, y)
DiscriminantAnalysis.ModelLDA{Float64}(4x4 Array{Float64,2}:
 0.788189  -1.25472   0.973844  -3.43881
 1.39847    3.45505   3.96094    2.80604
 0.669245  -1.04101  -0.575783   3.02589
 0.874092   1.62662  -2.69239   -1.97076,3x4 Array{Float64,2}:
 1.464  0.244  5.006  3.418
 4.26   1.326  5.936  2.77
 5.552  2.026  6.588  2.974,[0.3333333333333333,0.3333333333333333,0.33333333333
33333])
```

6. We can use the preceding fitted model to perform the classification and prediction tasks. This can be done using the `lda()` function with the datasets of the dependent and the independent variables as the input parameters, as follows:

```
pred = classify(model, X)
```

```
julia> pred = classify(model, X)
150x1 Array{Int64,2}:
 1
 1
 1
 1
 1
 1
 1
 1
 1
 1
 ⋮
 3
 3
 3
 3
 3
 3
 3
 3
```

7. Now, we can calculate the accuracy of the linear discriminant classification model to cross-check the performance of the classification done by our Linear Discriminant model. This can be done by dividing the sum of the matched terms (dependent variables) by the length of the dependent variable:

```
accuracy = sum(pred .== y) / length(y)
```

```
julia> accuracy = sum(pred .== y)/length(y)
0.98
```

How it works...

The `iris` dataset contains three classes or types of iris, each having 50 observations/instances. Every observation contains information such as the sepal width and length, petal width and length, and the class itself, which is the dependent variable.

As the dataset we are using is already in the CSV form at, the `readtable()` function does not require extra delimiter parameters to be added.

Pooling is the process that converts the different classes into factors. This helps the algorithm identify the different classes of the iris plant as different vectors, which helps in the classification process. As the `pool()` function changes the value of the data that it takes as an input, an extra ! symbol is added after it as per coding practices in Julia.

The `convert()` function is used to convert data to another type. So here, we are converting the dataset containing the independent variables into a common data type, which here is an Array of type `Float64`. We assign the variable X to the resultant dataset.

Now, we assign the variable y to the dataset containing the dependent variable, which is the column containing the **Species** details. The naming of the variables is done according to the general practices followed in the analytics community.

Finally, we use the `lda()` function of the `DiscriminantAnalysis` package to fit the Linear Discriminant Analysis model for classifying the species type in the dataset, depending on the values of the independent variables.

Now, we use the model fitted in the previous step to classify the data according to the values of the independent variables using the `classify()` function.

We can calculate the accuracy of the model we have fitted, and the classification done by it, by comparing the actual values and the predicted values of y. The difference between them is actually called the **error value**. There are a lot of different ways the error value is used to compute different types of errors, but understanding these are not required to understand the accuracy of the model we fitted for the classification task in the preceding section.

Data preprocessing

Data preprocessing is one of the most important parts of an analytics or a data science pipeline. It involves methods and techniques to sanitize the data being used, quick hacks for making the dataset easy to handle, and the elimination of unnecessary data to make it lightweight and efficient when used in the analytics process. For this recipe, we will use the `MLBase` package of Julia, which is known as the Swiss Army Knife of writing machine learning code. Installation and setup instructions for the library will be explained in the *Getting ready* section.

Getting ready

1. To get started with this recipe, you have to add the MLBase Julia package, which can be done by running the Pkg.add() function in the REPL. It can be done as follows:

```
Pkg.add("MLBase")
```

2. After installing the package, it can be imported using the using ... command in the REPL. It can be done as follows:

```
using MLBase
```

After importing the package following the preceding steps, you are ready to dive into the *How to do it...* section.

How to do it...

1. The first type of preprocessing technique that we will learn to use is data repetition. For repeating a value or a set of values multiple times, the repeach() function can be used. To repeat a set of values or elements for a fixed number of times for each, the repeach() function with the appropriate values, and the number repetitions, should be included as input parameters. It can be done as follows:

```
repeach(1:3, 2)
```

```
julia> repeach(1:3, 2)
6-element Array{Int64,1}:
 1
 1
 2
 2
 3
 3
```

2. The `repeach()` function can also be used for repeating ASCII strings. It can be done by including the characters to be repeated as the form of an array and including it as a parameter in the `repeach()` parameter along with the repetition value. This can be done as follows:

    repeach(["a", "b", "c"], 2)

```
julia> repeach(["a", "b", "c"], 2)
6-element Array{ASCIIString,1}:
 "a"
 "a"
 "b"
 "b"
 "c"
 "c"
```

3. The `repeach()` function can also be used for non-uniform repetition. In the preceding example, we repeated the observations for a fixed number of times for each. However, when the repetition elements are added as a list, we can also specify how many times a particular observation needs to be repeated. This provides more elegance and flexibility to the analyst using the function. This can be done as follows:

    repeach(["a", "b", "c"], [3, 2, 1])

```
julia> repeach(["a", "b", "c"], [3, 2, 1])
6-element Array{ASCIIString,1}:
 "a"
 "a"
 "a"
 "b"
 "b"
 "c"
```

4. To repeat every column in a particular matrix, we can use the `repeachcol()` function. This takes in the matrix and the repetition value as the input parameters and outputs a matrix with the repeated columns. This can be done as follows:

```
A = [2 -4 8.2; -5 3.5 63]
repeachcol(A, 2)
```

```
julia> A = [2 -4 8.2; -5.5 3.5 63]
2x3 Array{Float64,2}:
  2.0  -4.0   8.2
 -5.5   3.5  63.0

julia> repeachcol(A, 2)
2x6 Array{Float64,2}:
  2.0   2.0  -4.0  -4.0   8.2   8.2
 -5.5  -5.5   3.5   3.5  63.0  63.0
```

5. As done in the previous step, the `repeachcol()` function can also take an array of values as an input parameter for non-uniform repetition. This can be done as follows:

```
A = [2 -4 8.2; -5 3.5 63]
repeachcol(A, [1, 2])
```

```
julia> A = [2 -4 8.2; -5.5 3.5 63]
2x3 Array{Float64,2}:
  2.0  -4.0   8.2
 -5.5   3.5  63.0

julia> repeachcol(A, [1,2])
2x3 Array{Float64,2}:
  2.0  -4.0  -4.0
 -5.5   3.5   3.5
```

6. Similar to the previous step, the rows of a matrix can also be repeated using the `repeachrow()` function, which takes in the array and the repetition value as the input parameter and gives out an array as an output. This can be done as follows:

```
A = [2 -4 8.2; -5.5 3.5 63]
repeachrow(A, 2)
```

```
julia> A = [2 -4 8.2; -5.5 3.5 63]
2x3 Array{Float64,2}:
  2.0  -4.0   8.2
 -5.5   3.5  63.0

julia> repeachrow(A, 2)
4x3 Array{Float64,2}:
  2.0  -4.0   8.2
  2.0  -4.0   8.2
 -5.5   3.5  63.0
 -5.5   3.5  63.0
```

7. Similar to the column repetition function, the repeachrow() function also takes in an array of repetition values for non-uniform row repetition, as follows:

```
A = [2 -4 8.2; -5 3.5 63]
repeachrow(A, [1, 2])
```

```
julia> A = [2 -4 8.2; -5.5 3.5 63]
2x3 Array{Float64,2}:
  2.0  -4.0   8.2
 -5.5   3.5  63.0

julia> repeachrow(A, [1, 2])
3x3 Array{Float64,2}:
  2.0  -4.0   8.2
 -5.5   3.5  63.0
 -5.5   3.5  63.0
```

8. In classification problems in Machine Learning, we come across labels/classes for the observations. So, the MLBase package provides a LabelMap type that helps the analyst define a set of labels in an easier way. This can be done using the labelmap() function, which takes the set or the array of variables as an input argument. This can be done as follows:

```
labels = labelmap(["a", "b", "b", "c", "c", "c"])
```

```
julia> labels = labelmap(["a", "b", "b", "c", "c", "c"])
LabelMap (with 3 labels):
[1] a
[2] b
[3] c
```

9. After building a labelmap, the `labelencode()` function will help in creating and defining dummy values for the mapped variables. This would help in defining an integer value for the different classes, which makes the classification process easy and also makes the process of defining the threshold for classification easier. This can be done as follows:

```
labelencode(labels, ["b", "c", "c", "a"])
```

```
julia> labelencode(labels, ["b", "c", "c", "a"])
4-element Array{Int64,1}:
 2
 3
 3
 1
```

How it works...

Repetition of values in a dataframe is very common, especially, in the case of time series data. It is used for repeating names of days in the case of time series data. Even synthetic data contains a lot of repeated values in most cases. So, the function explained in Step 1 will help make creating repetitive data easier.

Similar to the preceding function, the data can also be repeated non-uniformly so that the task of creating custom synthetic data becomes very easy. This can be done by a different function, which is shown in Step 2 of the preceding section.

There are functions that also help in manipulating specific amounts or pieces of data. The function in Step 3 helps in repeating and creating the repetitive values of columns.

We have another similar function that we saw earlier. However, instead of columns, it helps in construct rows with repetitive values, as explained in Step 4.

The values of the dataset can be in the form of text classes or some ordinal type values. The function explained in the last step helps to handle that type of data. It does so by setting the classes for the ordinal text data, which basically dummifies the variables. It also sets ordinal classes for the data set if called on.

Linear regression

Linear Regression is a linear model that is used to determine and predict numerical values. Linear regression is one of the most basic and important starting points in understanding linear models and predictive analytics. For this recipe, we will use Julia's GLM.jl package.

Getting ready

To get started with this recipe, you have to add the GLM.jl Julia package. It can be added and imported in the REPL using the Pkg.add(" ") command just like we added other packages before. This can be done as follows:

```
Pkg.add("GLM")
```

Now, import the package using the using " " command. The DataFrames package is also required to be imported. This can be done as follows:

```
using GLM
using DataFrames
```

How to do it...

1. Here, we will attempt to perform a simple linear regression on two basic arrays, which we have generated on-the-fly. Let's call the two array A and B and then, create a dataframe containing them. This can be done as follows:

```
df = DataFrame(A = [3, 6, 9], B = [34, 56, 67])
```

```
julia> df = DataFrame(A = [3, 6, 9], B = [34, 56, 67])
3x2 DataFrames.DataFrame
| Row | A | B  |
|-----|---|----|
| 1   | 3 | 34 |
| 2   | 6 | 56 |
| 3   | 9 | 67 |
```

2. Now the ordinary least squares linear regression model can be fitted using the `glm()` function. It takes in `Normal()` and `IdentityLink()` as additional parameters, discussed later in the book. The regression formula is also included as a parameter in the `glm()` function. So this is how it can be done:

```
Reg = glm(B~A, df, Normal(), IdentityLink())
```

```
julia> Reg = glm(B~A, df, Normal(), IdentityLink())
DataFrames.DataFrameRegressionModel{GLM.GeneralizedLinearModel{GLM.GlmResp{Array
{Float64,1},Distributions.Normal,GLM.IdentityLink},GLM.DensePredChol{Float64,Bas
e.LinAlg.Cholesky{Float64,Array{Float64,2}}}},Float64}:

Coefficients:
             Estimate Std.Error z value Pr(>|z|)
(Intercept)   19.3333   6.85971 2.81839   0.0048
A                 5.5   1.05848 5.19615   <1e-6
```

3. Now, as we have fitted the regression model on the data, we can check the error and the predictions from the fit value. Let's start by checking the standard error value. It can be checked using the `stderr()` function, which takes in the fit as an input argument. This can be done as follows:

```
error = stderr(Reg)
```

```
julia> error = stderr(Reg)
2-element Array{Float64,1}:
 6.85971
 1.05848
```

4. The predictions from the fit can be calculated using the `predict()` function, which also takes the fit value as an input argument. This can be done as follows:

```
predict(Reg)
```

```
julia> prediction = GLM.predict(Reg)
3-element Array{Float64,1}:
 35.8333
 52.3333
 68.8333
```

5. Now, to find the variance covariance matrix for the fit, we can use the `vcov()` function, which takes in the regression fit as an input parameter. This can be done as follows:

```
vcov(Reg)
```

```
julia> vcov(Reg)
2x2 Array{Float64,2}:
  47.0556   -6.72222
  -6.72222   1.12037
```

How it works...

Dataframes are one of the most used and preferred data structures or formats for data by analysts. Dataframes facilitate quick data manipulation and analysis. So, multiple arrays can easily be converted into a single dataframe for future analysis.

The ordinary least squares regression is so called due to the error function used to compute the error, which is the square root of the squared differences between the truths dataset and the prediction dataset. So, the `glm()` function fits a regression by minimizing that error and thus determining the coefficients for the line equation.

The error can also be checked separately by the process underlined in Step 3. This gives the value of the error, which is determined as explained in the preceding point.

The predictions according to the thus fitted regression line can be viewed through the process underlined in Step 4. This function gives the final predicted values of the fit rather than intermediate fits used for the optimization of line coefficients.

The variance-covariance matrix is used to measure the way one variable is correlated to another, thus giving an idea about related variables, which also help the analyst in the construction of derived variables and makes it possible to create synthetic data.

Classification

Classification is one of the core concepts of data science and attempts to classify data into different classes or groups. A simple example of classification can be trying to classify a particular population of people as male and female, depending on the data provided. In this recipe, we will learn to perform score-based classification, where each class is assigned a score, and the class with the lowest or the highest score is selected depending on the problem and the analyst's choice.

Getting ready

To get ready, the MLBase library has to be installed and imported. So, as we already installed it for the *Preprocessing* recipe, we don't need to install it again. Instead, we can directly import it using the using MLBase command:

```
using MLBase
```

How to do it...

1. We will explore score-based classification algorithms and techniques by creating simple arrays and matrices that can fulfill our purpose. The first and the most important function is the classify() function, which takes in the input data and classifies the data by assigning scores, the distance, to each class. It also takes in an additional parameter, ord, short for ordering. The ord parameter takes in either of two values: Forward and Reverse. The default value is Forward. So, classification with default ordering can be done as follows:

```
classify([0.1,0.5,0.2])
```

```
julia> classify([0.1,0.5,0.2])
2
```

2. To use `Reverse` ordering, the parameter should be specified along with the dataset as an input argument for `classify`, as follows:

```
classify([0.1,0.5,0.2], Reverse)
```

```
[julia> classify([0.1,0.5,0.2], Reverse)
1
```

3. To get the labels along with the scores in the output, the `classify_withscore()` function can be used, which takes in the data and the ordering parameter as input arguments. This can be done as follows:

```
classify_withscore([0.1,0.5,0.2])
```

```
julia> classify_withscore([0.1,0.5,0.2])
(2,0.5)
```

4. To get the labels and scores, when the lowest score indicates the best match, we can use the `Reverse` ordering value as an additional input parameter:

```
classify_withscore([0.1,0.5,0.2], Reverse)
```

```
[julia> classify_withscore([0.1,0.5,0.2], Reverse)
(1,0.1)
```

5. Finally, we will look at the classification of matrices that contain several columns, so there will be multiple labels and label scores. The `classify_withscores()` function is used to calculate the scores and determine the labels, and it also takes the *ordering* value as an optional parameter. This can be done as follows:

```
classify_withscores([[0.2 0.5 0.3; 0.7 0.6 0.2]'])
```

```
[julia> classify_withscores([0.2 0.5 0.3; 0.7 0.6 0.2])
([2,2,1],[0.7,0.6,0.3])
```

6. Classification for reverse ordering can be done by adding the `Reverse` keyword to the ordering argument in the `classify_withscores()` function. This can be done as follows:

```
classify_withscores([0.2 0.5 0.3; 0.7 0.6 0.2], Reverse)
```

```
julia> classify_withscores([0.2 0.5 0.3; 0.7 0.6 0.2], Reverse)
([1,1,2],[0.2,0.5,0.2])
```

How it works...

This classification is called **score-based classification** as the classification is done by assigning scores to the observations according to their distance from the classes. So, as the order influences the score-assigning mechanism, the classification depends on the order specified in the function. So, the reverse order can also be specified in the function, which would allow the analyst to classify the dataset in the reverse order.

To get the score that has been assigned, as discussed in the preceding section, the `classify_withscore()` function can be used. This gives the number of classes and the scores assigned.

The preceding two functions are also applied to multivariate data, which is the case in most real-world classification applications. This will return vectors with a list of classes, and the scores assigned to them, respectively.

Performance evaluation and model selection

Analysis of performance is very important for any analytics and machine learning processes. It also helps in model selection. There are several evaluation metrics that can be leveraged on ML models. The technique depends on the type of data problem being handled, the algorithms used in the process, and also the way the analyst wants to gauge the success of the predictions or the results of the analytics process.

Getting ready

To get ready, the `MLBase` library has to be installed and imported. So, as we already installed it for the *Preprocessing* recipe, we don't need to install it again. Instead, we can directly import it using the `using MLBase` command.

How to do it...

1. Firstly, the predictions and the ground truths need to be defined in order to evaluate the accuracy and performance of a machine learning model or an algorithm. They can take a simple form of a Julia array. This is how they can be defined:

   ```
   truths = [1, 2, 2, 4, 4, 3, 3, 3, 1]
   pred   = [2, 2, 2, 4, 3, 3, 2, 3, 1]
   ```

   ```
   julia> truths = [1, 2, 2, 4, 4, 3, 3, 3, 1]
   9-element Array{Int64,1}:
    1
    2
    2
    4
    4
    3
    3
    3
    1

   julia> pred   = [2, 2, 2, 4, 3, 3, 2, 3, 1]
   9-element Array{Int64,1}:
    2
    2
    2
    4
    3
    3
    2
    3
    1
   ```

2. Now, to get the rate or percentage of correct predictions, we can use the `correctrate()` function, which takes in the ground truth and the prediction sets to determine the rate of accuracy. This can be done as follows:

   ```
   correctrate(truths, pred)
   ```

```
[julia> correctrate(truths, pred)
0.6666666666666666
```

3. Similar to the correctness calculation, the error rate can also be calculated. This would be the percentage of observations that have not been predicted correctly. The value, intuitively, would be the difference between 1 and the value of the correct rate, which we calculated previously. The error rate can be calculated using the errorrate() function, which takes in the ground truths and the prediction sets as input arguments. This can be done as follows:

```
errorrate(truths, pred)
```

```
julia> errorrate(truths, pred)
0.3333333333333333
```

4. The confusion matrix with respect to the truth and the predicted value sets can be calculated by specifying the number of classes. As an output, it gives an integer value that takes into account the number of elements matching the ground truths and the prediction sets and belonging to a particular class. It can be calculated through the confusmat() function, which takes in the number of classes, the ground truth set, and the prediction set. This can be done as follows:

```
confusmat(4, truths, pred)
```

```
julia> confusmat(4, truths, pred)
4x4 Array{Int64,2}:
 1  1  0  0
 0  2  0  0
 0  1  2  0
 0  0  1  1
```

5. **Receiver Operating Characteristics (ROC)** are used to calculate the performance of a Machine Learning algorithm, specifically the classification algorithms. An ROC model can be fit using the roc() function, which takes in the ground truths and the prediction matrices. This can be done as follows:

```
r = roc(truths, pred)
```

```
julia> roc(truths, pred)
MLBase.ROCNums{Int64}
  p = 9
  n = 0
  tp = 6
  tn = 0
  fp = 0
  fn = 0
```

6. As we have fitted an ROC curve in the preceding step, we can now calculate the various performance measurements we can get through the fit, such as true positives, false positives, and so on.

7. The true positive and true negative values can be measured through the `true_positive()` and `true_negative()` functions, which take in the ROC curve fit as an input parameter. This can be done as follows:

```
true_positive(r)
true_negative(r)
```

```
julia> true_positive(r)
6

julia> true_negative(r)
0
```

8. Similarly, the false positive and false negative rates can also be calculated. The recall value can be calculated using the `recall(r)` function, which takes in the ROC curve fit, `r`, as an input argument. It is the true positive rate. This can be done as follows:

```
recall(r)
```

```
julia> recall(r)
0.6666666666666666
```

How it works...

The prediction and the ground truth datasets are absolutely necessary for determining the model performance and accuracy. The predictions dataset is, as the name suggests, a dataset of the prediction values, and the truths dataset contains the actual values. So, the accuracy and precision of the model and the algorithm can be calculated through the difference between the two datasets.

The `correctrate()` function gives the number of observations in the predictions dataset matching the corresponding observations in the truths dataset. This metric helps in cross-evaluating the accuracy of the model.

The `errorrate()` function is basically the reverse of the `correctrate()` function. It gives the number of predictions that differ from the corresponding truths dataset. This metric helps in analyzing the reliability of the model and the algorithm.

A confusion matrix is a great way to understand the performance of an algorithm or a technique. It is a table or a matrix where the true values and the predictions are plotted against. It will be an *nXn* matrix, *n* being the length of either of the datasets.

The ROC is a plot that plots the values of **true positive rate** versus the **false positive rate** values for the results of a classifier. The false positive rate is called the **fall out**, whereas the true positive rate is called **sensitivity**. ROC are generally used for setting or calculating the threshold values. The area under the ROC curve ranks the probability of a randomly chosen positive instance higher than a randomly chosen negative instance.

The true positive and the true negative metrics can be calculated using the custom functions for them, namely `true_positive()` and `true_negative()`. The recall value, which is another word for sensitivity, can also be calculated using the `recall()` function.

Cross validation

Cross validation is one of the most underrated processes in the domain of data science and analytics. However, it is very popular among the practitioners of competitive data science. It is a model evaluation method. It can give the analyst an idea about how well the model would perform on new predictions that the model has not yet seen. It is also extensively used to gauge and avoid the problem of overfitting, which occurs due to an excessive precise fit on the training set leading to inaccurate or high-error predictions on the testing set.

Getting ready

To get ready, the MLBase library has to be installed and imported. So, as we already installed it for the *Preprocessing* recipe, we don't need to install it again. Instead, we can directly import it using the using MLBase command. This can be done as follows:

```
using MLBase
```

How to do it...

1. Firstly, we will look at the k-fold cross-validation method, which is one of the most popular cross validation methods used. The input data set is randomly sampled into several sets. This can be done using the Kfold() function, which takes in the training dataset and the value of k, which is the number of validation sets that need to be produced. Here, we will generate an array of 10 integers for a quick example. The collect function is used over Kfold() to collect the sets created by it. Go ahead and try it without the collect() function too. This can be done as follows:

```
collect(Kfold(10, 3))
```

```
[julia> collect(Kfold(10, 3))
3-element Array{Any,1}:
 [1,2,6,7,8,9,10]
 [2,3,4,5,8,9]
 [1,3,4,5,6,7,10]
```

2. The next method of cross validation is leave-out-one cross validation. This technique creates multiple cross validation sets with the size (*n-1*), where *n* is the length of the dataset we want to perform cross validation upon. The LOOCV() function can be used for this technique. It takes in the length of the element array as an input parameter. This can be done as follows:

```
collect(LOOCV(n))
```

```
[julia> x = [1,2,3,4,5,6,7,8,9]
9-element Array{Int64,1}:
 1
 2
 3
 4
 5
 6
 7
 8
 9

[julia> collect(LOOCV(8))
8-element Array{Any,1}:
 [2,3,4,5,6,7,8]
 [1,3,4,5,6,7,8]
 [1,2,4,5,6,7,8]
 [1,2,3,5,6,7,8]
 [1,2,3,4,6,7,8]
 [1,2,3,4,5,7,8]
 [1,2,3,4,5,6,8]
```

3. Randomly subsampling is another cross validation technique, where the dataset is repeatedly subsampled randomly to generate cross validation subsets. This can be done using the RandomSub() function, which takes in the dataset, the length of the samples, and the number of samples needed. This can be done as follows:

```
RandomSub(5, 5, 4)
```

```
[julia> collect(RandomSub(5, 5, 4))
4-element Array{Any,1}:
 [1,2,3,4,5]
 [1,2,3,4,5]
 [1,2,3,4,5]
 [1,2,3,4,5]
```

How it works...

The k-fold cross validation technique is one of the most widely used cross validation methods by statisticians. In this method, the cross validation dataset is sampled into k subsamples, and the model or the approach of the analyst is tested on those subsampled datasets. This helps in tackling the problem of overfitting.

The **leave-out-one cross validation** (**LOOCV**) method, as the name suggests, creates sample sets that consist of all but one of the elements of the cross validation dataset. This is also one of the cross validation methods that help in tackling fit problems of the models and algorithms. The difference between this and the preceding method is that this has a higher overlap with the test sets, as the k-fold method creates non-overlapping subsets.

The Random Subsample function has the same purpose as the preceding techniques. However, it is completely random, and the overlap rate cannot be determined or controlled. So, it is more useful for uncontrolled, random cross-validation analysis.

Distances

In statistics, the distance between vectors or data sets are computed in various ways depending on the problem statement and the properties of the data. These distances are often used in algorithms and techniques such as recommender systems, which help e-commerce companies such as Amazon, eBay, and so on, to recommend relevant products to the customers.

Getting ready

To get ready, the `Distances` library has to be installed and imported. We install it using the `Pkg.add()` function. It can be done as follows:

```
Pkg.add("Distances")
```

Then, the package has to be imported for use in the session. It can be imported through the `using ...` command. This can be done as follows:

```
using Distances
```

How to do it...

1. Firstly, we will look at the Euclidean distance. It is the ordinary distance between two points in Euclidean space. This can be calculated through the Pythagorean distance calculation method, which is the square root of the square of the element-wise differences. This can be done using the `evaluate()` function, which takes in the distance metric along with the datasets between which we want to find the distance. This can be done as follows:

```
        d = evaluate(Euclidean(), a, b)
```

```
julia> a = [1, 2, 3, 4]
4-element Array{Int64,1}:
 1
 2
 3
 4

julia> b = [4, 3, 5, 1]
4-element Array{Int64,1}:
 4
 3
 5
 1

julia> evaluate(Euclidean(), a, b)
4.795831523312719
```

2. We can also evaluate the distance between the columns in a dataset.
   The colwise() function takes in the distance type and the matrices whose
   column-wise distance we intend to find. This can be done as follows:

```
        colwise(dist, a, b)
```

```
julia> a = [1 2 3; 4 5 6]
2x3 Array{Int64,2}:
 1  2  3
 4  5  6

julia> b = [7 8 9; 10 11 12]
2x3 Array{Int64,2}:
  7   8   9
 10  11  12

julia> colwise(Euclidean(), a, b)
3-element Array{Float64,1}:
 8.48528
 8.48528
 8.48528
```

3. The distances between each pair of columns can also be computed. This helps in getting the distance between two desired independent variables in the data. The `pairwise()` function takes in the distance type and the matrices whose pairwise distances we are intending to calculate. This can be done as follows:

```
pairwise(dist, a, b)
```

```
julia> pairwise(Euclidean(), a, b)
3x3 Array{Float64,2}:
 6.7082    9.89949   11.3137
 5.38516   8.48528    9.89949
 4.12311   7.07107    8.48528
```

4. The same `pairwise()` function can also be used to compute the distances between the columns of the same matrix. In this case, the function takes in just the distance metric and the matrix as the input arguments. This can be done as follows:

```
julia> pairwise(Euclidean(), a)
3x3 Array{Float64,2}:
 0.0      1.41421   2.82843
 1.41421  0.0       1.41421
 2.82843  1.41421   0.0
```

5. There is a very diverse list of distance metrics in the `Distances.jl` package, which also includes distances such as the `Hamming` distance, which is widely used to calculate the distance between binary strings. This can be done by substituting `Euclidean()` in the preceding examples with `Hamming()`, as follows:

```
evaluate(Hamming(), a, b)
```

```
julia> a
5-element Array{Int64,1}:
 2
 4
 6
 1
 4

julia> b
5-element Array{Int64,1}:
 4
 3
 6
 2
 1

julia> evaluate(Hamming(), a, b)
4
```

How it works...

The evaluate function simply takes in the two datasets or arrays whose distances need to be calculated and the distance type for evaluation. There are multiple distance types including the Euclidean and Minkowski distances, which are commonly used in recommender systems, and the cosine similarity distance, which is also sometimes used in recommender systems in the case of sparsely distributed data.

The distance between the two columns of a dataset gives the distance or the relationship between two independent variables of a dataset, for example, a dataset of rainfall for all the states.

The pairwise distance of all columns can be computed using the pairwise() function. This helps by having a single formula for the distance computations rather than computing the distances for all the observations separately.

The pairwise() function that we used in the preceding example can also be used to compute the distances between the columns in the same matrix in addition to computing the distance between columns of two different matrices. This helps in situations flagged in Step 2 by enabling quick evaluation instead of writing long for-loops.

The Distances.jl package has an exhaustively long list of distance metrics that can be calculated. Each distance plays an important role in each different situation, as explained in the Step 1. The distance metric selection depends on the problem statement and the properties of the data.

Distributions

A probability distribution is when each point or subset in a randomized experiment is allotted a certain probability. So, every random experiment (and, in fact, the data of every data science experiment) follows a certain probability distribution. And the type of distribution being followed by the data is very important for initiating the analytics process, as well as for selecting the machine learning algorithms that are to be implemented. It should also be noted that, in a multivariate data set, each variable might follow a separate distribution. So, it is not necessary that all variables in a dataset follow similar distributions.

Getting ready

To get ready, the `Distributions` library has to be installed and imported. We install it using the `Pkg.add()` function, as follows:

```
Pkg.add("Distributions")
```

Then the package has to be imported for use in the session. It can be imported through the `using ...` command, as follows:

```
using Distributions
```

How to do it...

1. Firstly, let's start by understanding how to work with a normal distribution, which is the most popular and also the most important distribution for analysts. So we start off by creating a random seed through the `srand()` function, then we generate a normal distribution through the `Normal()` function, and finally we draw samples from the thus created normal distribution using the `rand()` function, which takes in the distribution variable and the desired number of samples as input arguments. This can be done as follows:

```
srand(110)
n = Normal()
s = rand(n, 10)
```

```
julia> srand(110)
MersenneTwister(Base.dSFMT.DSFMT_state(Int32[1178870109,1072829093,579093040,107
3524090,311037256,1073165746,-2073743796,1073717721,1401677141,1073436599  …  21
09078678,1072795008,1043120363,1073006388,-2126404085,1582055288,-1953405793,109
0546129,382,32700]),[1.76845,1.94052,1.67396,1.39545,1.31324,1.66255,1.58602,1.0
5213,1.26864,1.10887  …  1.37456,1.51655,1.82701,1.13072,1.05724,1.76705,1.16378
,1.59908,1.46599,1.18902],382,UInt32[0x0000006e])

julia> n = Normal()
Distributions.Normal(μ=0.0, σ=1.0)

julia> s = rand(n, 10)
10-element Array{Float64,1}:
 -0.285446
  0.265386
 -1.10066
  0.667889
  0.740703
 -0.637788
 -0.26416
  2.64611
 -0.166272
  0.302357
```

2. Statistics such as the probability density function (pdf), the cumulative distributive function, and the percentile values can be calculated using the pdf, cdf, and percentile functions, respectively. The cdf() function takes in the distribution variable and an integer variable to give us the cumulative probability distribution value. The pdf() function takes in the distribution and a float value to give the probability distribution function. The quantile() function takes in the distribution variable and a set of quantile values where the distribution value is intended to be calculated. This can be done as follows:

```
pdf(n, 0.4)
cdf(n, 3)
quantile(n, [0.25, 0.5, 0.95])
```

```
julia> pdf(n, 0.4)
0.36827014030332333

julia> cdf(n, 3)
0.9986501019683699

julia> quantile(n, [0.25, 0.5, 0.95])
3-element Array{Float64,1}:
 -0.67449
  0.0
  1.64485
```

3. Finally, as the normal distribution generated in the preceding example is parameterized by the mean and the standard deviation, we can draw random samples from it for experimentation or testing purposes or simply to use them as data points in an experiment. It can be done using the `rand()` function, which takes in the distribution construction and the number of samples needed. This can be done as follows:

```
rand(Normal(1, 2), 100)
```

```
julia> rand(Normal(0, 1), 10)
10-element Array{Float64,1}:
  0.651388
 -0.552412
  0.423087
 -0.756068
  0.0294799
 -0.558174
 -1.22044
  0.316617
 -0.00874207
 -0.492949
```

4. Using the `fit()` function, which takes in the distribution name and an array, we can know the parameters of the closest distribution it follows of the type we entered in the `fit()` function arguments. This can be done as follows:

```
fit(Normal, rand(10))
```

```
julia> fit(Normal, rand(10))
Distributions.Normal(μ=0.47444700737751055, σ=0.26805235118694953)
```

5. The `Distributions` package also allows constructing mixture models. This can be done very easily by defining the `MixtureModel()` function, which takes in the distribution type along with the set of distributions in the form of an array and an optional argument that takes in an array of priors for each of the distributions. This can be done as follows:

```
MixtureModel(Normal[Normal(2.0, 1.2), Normal(0.0, 1.0),
Normal(3.0, 1.5)])
```

```
julia> MixtureModel(Normal[
                Normal(2.0, 1.2),
                Normal(0.0, 1.0),
                Normal(3.0, 1.5)]
            )
MixtureModel{Distributions.Normal}(K = 3)
components[1] (prior = 0.3333): Distributions.Normal(μ=2.0, σ=1.2)
components[2] (prior = 0.3333): Distributions.Normal(μ=0.0, σ=1.0)
components[3] (prior = 0.3333): Distributions.Normal(μ=3.0, σ=1.5)
```

How it works...

The Normal() function creates a normal distribution with the specified number of observations with the mean and the standard deviation 1, which are the default values. The number of observations should also be included in the function. The seed random function, srand(), just ensures deterministic values of the normal distribution every time it is used so that the values aren't different when used in different operations.

The probability density function and the cumulative distribution functions of the normal distributions can be calculated by the pdf and cdf functions. The pdf value gives the probability of a particular random variable taking a specific value. The cumulative distribution function gives the cumulative probability value for a random variable taking a value less than or greater than a specific value. The quantiles of a distribution give the values of the 25%, 50%, and 75% of the normal distribution.

However, the distributions can also be constructed with user-defined mean and standard deviation values, and then a specified number of values can be parameterized from the thus constructed distribution, which makes a good random sample for analytics purposes.

Whenever we want to know what type of distribution the data is following, we can take that data in the form of an array and then use the fit() function with the distribution in mind. This gives the spread and fit of the distribution with respect to the data, thus helping the analyst to choose a specific model for the fitted distribution type.

Mixture models of multiple distributions can also be constructed through the Mixture() function. This is especially handy when dealing with a mixture of multiple datasets, each following a distribution with different parameter values.

Time series analysis

Time series is another very important form of data. It is more widely used in stock markets, market analysis, and signal processing. The data has a time dimension, which makes it look like a signal. So, in most cases, signal analysis techniques and formulae are applicable for time series data, such as autocorrelation, crosscorrelation, and so on, which we have already dealt with in the previous chapters. In this recipe, we will deal with methods to get around and work with datasets with the time series format.

Getting ready

To get ready for the recipe, the `TimeSeries` and `MarketData` libraries have to be installed and imported. We install them using the `Pkg.add()` function, as follows:

```
Pkg.add("TimeSeries")
Pkg.add("MarketData")
```

Then the package has to be imported for use in the session. It can be imported through the `using ...` command, as follows:

```
using TimeSeries
using MarketData
```

How to do it...

1. The `TimeArray` format from the `TimeSeries` package makes it easy to interpret and work with the time-series data format. Let's construct a `TimeArray` by constructing an array of elements of the type `Date`. It can be created using the `Date()` constructor, as follows:

```
dates = [Date(2009, 1, 1) : Date(2010, 12, 31)]
```

```
julia> dates = [Date(2009, 1, 1) : Date(2010, 12, 31)]
730-element Array{Date,1}:
 2009-01-01
 2009-01-02
 2009-01-03
 2009-01-04
 2009-01-05
 2009-01-06
 2009-01-07
 2009-01-08
 2009-01-09
 2009-01-10
 ⋮
 2010-12-23
 2010-12-24
 2010-12-25
 2010-12-26
 2010-12-27
 2010-12-28
 2010-12-29
 2010-12-30
 2010-12-31
```

2. Now we can use the `TimeArray()` constructer to build the TimeArray from the array of dates we have just created. The constructor takes in the `dates` array and a set of values that can be assigned according to the date dimension. We will generate those values randomly with the length of the `dates` array. This can be done as follows:

```
times = TimeArray(dates, rand(length(dates)), ["random column"])
```

```
julia> times = TimeArray(dates, rand(length(dates)), ["random column"])
730x1 TimeSeries.TimeArray{Float64,1,Date,Array{Float64,1}} 2009-01-01 to 2010-1
2-31

              random column
2009-01-01 | 0.9542
2009-01-02 | 0.5886
2009-01-03 | 0.8523
2009-01-04 | 0.8232
 ⋮
2010-12-28 | 0.1693
2010-12-29 | 0.3096
2010-12-30 | 0.1438
2010-12-31 | 0.544
```

3. Array indexing can be done just like in a normal array with the range defined in the square brackets after the name of the array, as follows:

```
times[3 : 7]
```

```
julia> times[3 : 7]
5x1 TimeSeries.TimeArray{Float64,1,Date,Array{Float64,1}} 2009-01-03 to 2009-01-
07

            random column
2009-01-03 |  0.8523
2009-01-04 |  0.8232
2009-01-05 |  0.3853
2009-01-06 |  0.3571
2009-01-07 |  0.6353
```

4. In addition to length indexing, column-wise indexing can also be done just like dataframes in Python and R, by entering the column name in the square brackets instead of the index range, as follows:

```
times["random column"]
```

```
julia> times["random column"]
730x1 TimeSeries.TimeArray{Float64,1,Date,Array{Float64,1}} 2009-01-01 to 2010-1
2-31

            random column
2009-01-01 |  0.9542
2009-01-02 |  0.5886
2009-01-03 |  0.8523
2009-01-04 |  0.8232
⋮
2010-12-28 |  0.1693
2010-12-29 |  0.3096
2010-12-30 |  0.1438
2010-12-31 |  0.544
```

5. The `from` keyword provides a simple and elegant way to do time series indexing according to a specific period. There is also a `to` keyword, which does exactly the opposite of what `from` does. Both take in the `TimeArray` and the dates from/to which the time series need to be indexed. This can be done as follows:

```
from(times, Date(2009, 8, 30))
```

```
[julia> from(times, Date(2009, 8, 30))
489x1 TimeSeries.TimeArray{Float64,1,Date,Array{Float64,1}} 2009-08-30 to 2010-1
2-31

               random column
2009-08-30 | 0.9028
2009-08-31 | 0.7878
2009-09-01 | 0.5906
2009-09-02 | 0.795
⋮
2010-12-28 | 0.1693
2010-12-29 | 0.3096
2010-12-30 | 0.1438
2010-12-31 | 0.544
```

```
to(times, Date(2009, 3, 1))
```

```
[julia> to(times, Date(2009, 3, 1))
60x1 TimeSeries.TimeArray{Float64,1,Date,Array{Float64,1}} 2009-01-01 to 2009-03
-01

               random column
2009-01-01 | 0.9542
2009-01-02 | 0.5886
2009-01-03 | 0.8523
2009-01-04 | 0.8232
⋮
2009-02-26 | 0.9198
2009-02-27 | 0.8062
2009-02-28 | 0.5094
2009-03-01 | 0.2336
```

6. The lag and lead values of the time series can be calculated by simply using the lag() and lead() functions. They take in the time series array, both as a whole or an indexed version of the time series array. This can be done as follows:

```
lag(times[3 : 7])
```

```
julia> lag(times[3:7])
4x1 TimeSeries.TimeArray{Float64,1,Date,Array{Float64,1}} 2009-01-04 to 2009-01-
07

            random column
2009-01-04 |  0.8523
2009-01-05 |  0.8232
2009-01-06 |  0.3853
2009-01-07 |  0.3571
```

```
lead(times[3 : 7])
```

```
julia> lead(times[3:7])
4x1 TimeSeries.TimeArray{Float64,1,Date,Array{Float64,1}} 2009-01-03 to 2009-01-
06

            random column
2009-01-03 |  0.8232
2009-01-04 |  0.3853
2009-01-05 |  0.3571
2009-01-06 |  0.6353
```

7. Moving average values are a great way to smoothen the time series through considering a sliding window view. The number of values for calculating the moving average can be specified in the moving() function along with the type of metric and the time series array. This can be done as follows:

```
moving(times, mean, 10)
```

```
julia> moving(times, mean, 10)
721x1 TimeSeries.TimeArray{Float64,1,Date,Array{Float64,1}} 2009-01-10 to 2010-1
2-31

            random column
2009-01-10 |  0.5863
2009-01-11 |  0.5705
2009-01-12 |  0.5776
2009-01-13 |  0.4995
⋮
2010-12-28 |  0.5373
2010-12-29 |  0.5056
2010-12-30 |  0.4259
2010-12-31 |  0.4475
```

How it works...

The `TimeArray` type helps in converting a normal array with the date format elements into an array that looks more like a time series that is governed by the time dimension. This helps us implement R-or Pandas-like time-series implementations and algorithms on the data. So, in the first example, we created artificial data randomly, which has a time dimension that has already been created in the first step by converting the normal array into a TimeArray.

Just like indexing is done in normal arrays, TimeArrays also offer a flexible indexing style, both row-wise and column-wise. Row-wise indexing can be done by specifying the range of the rows you want to index. Column wise indexing is done by specifying the names of the columns in the brackets. This makes it easy to manipulate specific parts of the dataset, and also in ML algorithms where specific parts of the dataset are needed as an input.

As already pointed out, time series data behaves like a signal. So, almost all the signal processing algorithms are relevant in time series analysis problems. The from and to functions also help index and slice the time series according to the governing dimension, which is the time dimension.

The `lag` and `lead` values of time series are important to calculate, especially when analyzing the behavior of the time series and also while checking for correlation and/or correlated segments. So, this has been done easily using the `lag()` and `lead()` function implementations by specifying the number of segments we want to calculate or analyze.

Moving average analysis is a model where a moving window is used to calculate the average of a number of adjacent points and then use that value to set the value at that particular point. This helps in smoothing the time series and reducing the complexity of some complex time series data. For example, in the last example, the mean of 10 adjacent values would be calculated to set the value at a particular point.

5

Working with Visualizations

In this chapter, we will cover the following recipes:

- Plotting basic arrays
- Plotting dataframes
- Exploratory data analysis through plots
- Line plots
- Scatter plots
- Histograms
- Aesthetic customizations

Introduction

In this chapter, you will learn how to visualize and present data and analyze the findings from the data science approach you have adopted to solve a particular problem. There are various types of visualization to display your findings: bar plots, the scatter plots, pie charts, and so on, and it is very important to choose an appropriate method that can reflect your findings and work in a sensible and an aesthetically pleasing manner.

Importance of visualizations and reporting in data science:

Visualization is the art of displaying quantitative information in a sensible, legible, and aesthetically pleasing way. It consists of plotting quantitative information in the form of various graphs as well as putting forward or compiling the analyses and the results in a precise and a legible report.

Visualizations and reporting should always be done in such a way that the person or the group to whom they are being presented to should be able to follow and appreciate it with minimal background and difficulty. The selection of the plot or the visualization is extremely important in order to make the report clearer and avoid confusion. Sometimes, using a pie chart instead of a box plot or a scatter plot would result in a skewed display, which can affect the perception of the reader or whoever the report is being presented to.

So, visualizations play a very important role when dealing with complex data with several variables and several analyses. Graphs and plots help simplify complex information and understand the report or the analysis in a better way when it is visually presented. So, visualizing meaningful data in a meaningful way is the goal of a good data visualization or a good analysis/data science report.

Plotting basic arrays

Arrays are one of the fundamental data structures used in data analysis to store various types of data. They are also a quick way to store columns or dimensions in data, for statistical analysis as well as exploratory analysis through plots and visualization. Arrays are also very easy to plot, as they are simple. When a visualization is being done with two columns of a dataset, it means that the two column values are taken in the form of separate arrays and then plotted against each other, which again makes arrays very important.

Getting ready

To get started with this recipe, you have to install the Gadfly library. This can be done using the following command:

```
Pkg.add("Gadfly")
```

Next, to import the library, we can import it by calling by its name, which is *Gadfly*. This can be done as follows:

```
using Gadfly
```

How to do it...

For this recipe, you need to perform the following steps:

1. Firstly, let's generate two random arrays *a* and *b* and plot them against each other. We can use the `rand` function to generate an array of random numbers, by giving it the length of the array as an input argument. This can be done as follows:

```
plot(x = rand(10), y = rand(10))
```

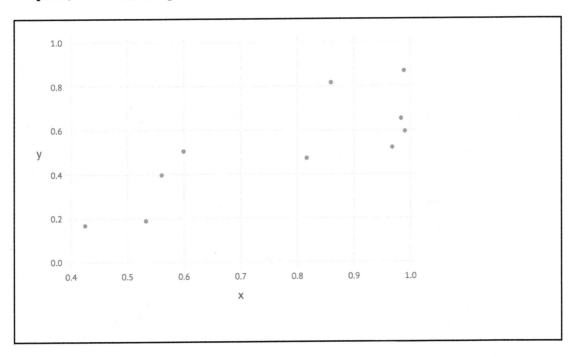

2. As there are no plot aesthetic elements defined, a point geometry is assumed by default.
3. The scatter plot can be further improved by connecting the points with lines so that we can get a broken line plot. This can be done by adding in the `Geom.line` argument in the preceding `plot()` function. We will plot two random arrays against each other, just like we did in the previous example. This can be done as follows:

```
plot(x = rand(10), y = rand(10), Geom.point, Geom.line)
```

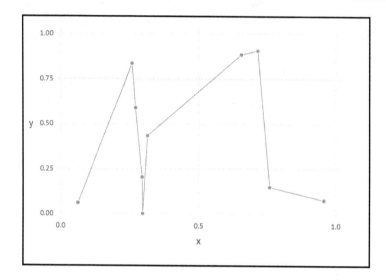

4. Gadfly scatter plots can also be customized to scale according to the values in either of the axis. This helps in plotting very high and very low values without distorting the plot. This can be done by including the `Scale` argument in the `plot()` function along with the axis and the type of scaling needed, as follows:

```
plot(x = rand(10), y = 2.^rand(10), Scale.y_sqrt, Geom.point, Geom.smooth)
```

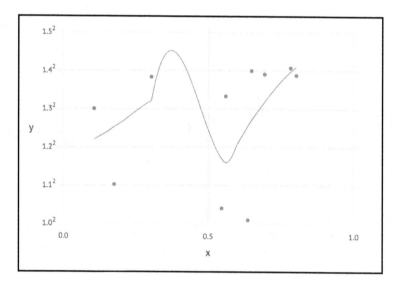

5. The plot aesthetics and legends can also be added in the same `plot()` function as additional arguments. The Gadfly `plot()` function takes in the `Guide` argument as the legend details, which include the labels for the *x* and *y* axes and also the plot title. This can be done as follows:

```
plot(x = rand(10), y = rand(10), Guide.xlabel("X-label"), Guide.ylabel("Y-label"), Guide.title("Plot title"))
```

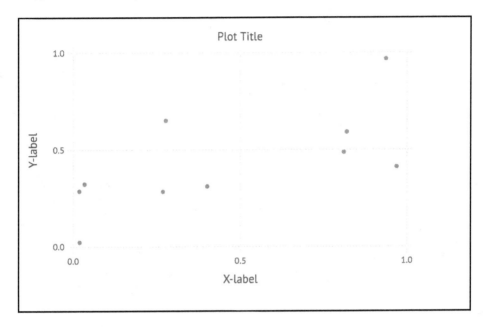

How it works...

A point geometry plot is popularly called a **scatter plot**. It takes in two arrays as its input aesthetics. This scatter plot is a bare skeleton when it comes to aesthetics. In other words, it is the simplest scatter plot with no added Gadfly aesthetics.

Adding aesthetics adds value to the plot and displays quantitative information better and in a more informed way. Adding the `geom_line` argument to the `plot()` function converts the scatter plot into a broken line plot, which helps us understand the data trends and the spread of the data through the `Geom_point` argument in the `plot()` function.

When data points from two different scales are plotted against each other, the uneven nature of the numbers might break the plot. So, the `scale` arguments help us deal with this issue. The values are represented as squares or powers according to the argument on the plot, which helps preserve the structure of the plot and represents the values better.

Plot legends and aesthetics are always helpful, and they add a lot of value to the plot. Aesthetics such as the *x* label, the *y* label, and other details help us understand and make sense of the data being displayed in the plot. It is recommended aesthetics that are as simple as possible so that they don't dominate the visual information on the plot.

Plotting dataframes

Dataframes are one of the datastructures on which most analytics and machine learning implementations are done. It is the most popular and best way for representing tabular data. They are made up of several arrays and similar data structures, and they can store data in multiple formats, including logical data, string data, and numeric data. So, visualizations can be done against one or multiple columns of the same dataframe, which makes it easy for the analyst to express numerical information in the dataframe.

Getting ready

To get started with this recipe, you have to install the `Gadfly` library as you did in the previous recipe.

As we will be using the datasets from **R** packages, we also need to import the `RDatasets` package. This can be done simply by the `using` ... syntax, which we use for importing packages:

```
using RDatasets
```

How to do it...

For this recipe, you need to perform the following steps:

1. Firstly, we will learn how to plot different columns of a dataframe against each other as a scatter plot. This is a very quick and efficient way to learn about any possible correlations between the different dimensions or columns of the dataset. This can be done just like we did for the arrays in the preceding section. The dataset can be loaded though the `dataset()` argument inside the `plot()`

function. So, let's now plot the Sepal width against the Sepal length of the `iris` dataset from the `RDatasets` package. This can be done as follows:

```
plot(dataset("datasets", "iris"), x = "SepalWidth", y = "SepalLength",
Geom.point)
```

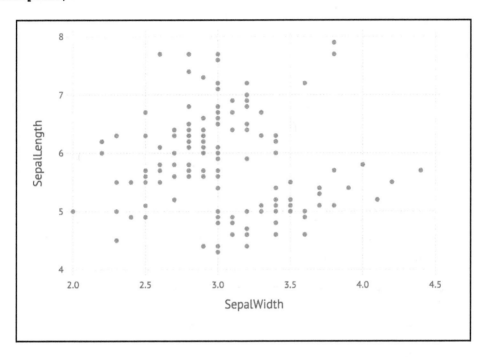

2. Dataframes containing multiple classes can be plotted as a series of stacked histograms with separate colors for each class. Stacked charts are very efficient in displaying quantitative information containing multiple classes. This can be done by adding in a `color` argument in the `plot()` function, which points to the column containing the class names or the corresponding logical values. Let's plot a stacked histogram of the Sepal width of the `iris` dataset, with a different color for each species. This can be done as follows:

```
plot(dataset("datasets", "iris"), x = "SepalWidth", color = "Species",
Geom.point)
```

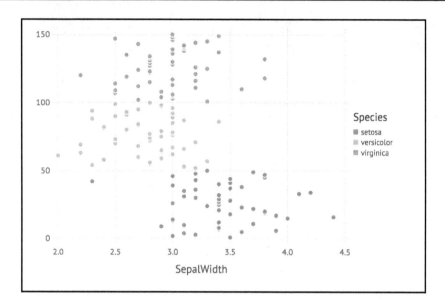

3. To get a stacked bar chart, which helps you understand frequencies and numbers of different categories in the dataset, the `Geom.histogram` parameter can be used in the `plot()` function, as follows:

```
plot(dataset("car", "SLID"), x = "Wages", color = "Language",
Geom.histogram)
```

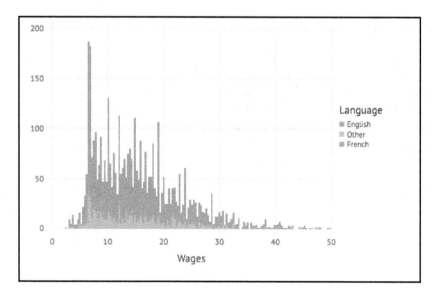

How it works…

Scatter plots are a basic and good way to get an idea about correlations and any other possible relationship between variables in a dataframe. The plot, here, plots a scatter plot between the width and the length of the iris flower. We can see a kind of a linear relationship between both the columns/features.

We can further improve the preceding plot we got using aesthetics. We can further segment that plot by allotting different colors to different types of flower by adding the *color* aesthetic. This helps us in further exploratory analytics on the relationship between the petal size and the petal width. So, we can now see that the virginica species has larger petals, as compared to the other two.

Another plot is the histogram plot, which basically plots the frequency distribution of a particular feature in a dataset. In this example, we have plotted the distribution of the wages of a particular demographic, which is the **Survey of Labour and Income Dynamics (SLID)** dataset. So, from this data, we can see that the hourly wage lies between the $5-$20 range for most workers. Now, by further applying the grouping color aesthetics, we can also see that the English workers tend to earn more than the French and the other demographic workers.

Plotting functions

In data science and statistical modeling, there are several instances where an analyst needs to use several functions for both transforming and exploratory analytics steps. So, one can plot them in Gadfly in a very simple way, which can used to plot separate functions as well as to stack several functions in a single plot.

Getting ready

As we already specified, we will use the Gadfly plotting library for this recipe too. So, follow the installation steps from the previous recipes.

How to do it...

1. Let's start with a basic function plot to get familiar with the syntax. So, a good basic function to start is the `sin()` function, which can be invoked as *sin*. The function can be included directly in the plot command, along with the upper and lower limits of the x axis. The syntax is: `plot(function, lower_limt, upper_limit)`. This can be done as follows:

```
plot(sin, 0, 30)
```

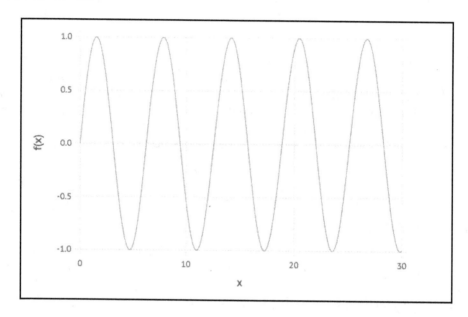

2. Similarly, if we want to plot multiple functions on a single plot, we can do just like we did in the previous example. Now, instead of including a simple function in the `plot` command, we have to include multiple functions. This can be achieved by adding them to an array inside the `plot` command. The syntax looks like this: `plot([function1, function2], lower_limit, upper_limit)`. Each function's plot takes a different color for itself so that the analyst doesn't need to worry a lot about aesthetics. This can be done as follows:

```
plot([sin, cos], 0, 30)
```

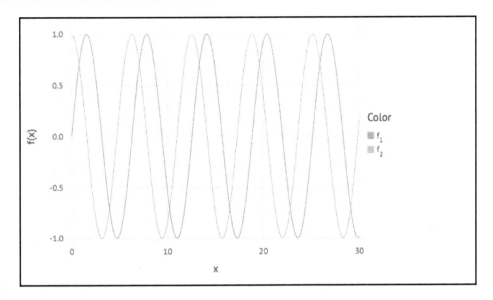

3. The functions can also be plotted as different layers in the same plot so that each function's aesthetics and style can be customized according to convenience. The layers can help distinguish data, which follows a very similar distribution. The syntax of layering in the plot looks like this:

```
plot(layer(x = ..., y = ..., aesthetics customization), layer(x = ..., y = ..., aesthetics customization))
```

4. The plot styles and color can also be varied. This can be done as follows:

```
plot(layer(x = rand(10), y = rand(10), Geom.line), layer(x = rand(10), y = rand(10), Geom.point))
```

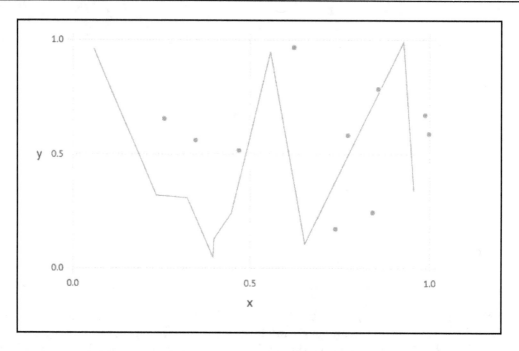

5. Legends for the plot can be added using the `Guide` parameter inside the `plot()` function. The syntax is very similar to normal plots, and it looks like this: `plot(layer(), layer(), Guide.XLabel("...")`, `Guide.YLabel("...")`, `"Guide.Title("..."))`. The *x* and *y* labels can be added inside the corresponding calls and the title of the plot in the `Guide.Title()` call. This can be done by generating random synthetic data, as follows:

```
plot(layer(x = rand(10), y = rand(10), Geom.point), layer(x = rand(10), y =
rand(10), Geom.line), Guide.XLabel("x-label"), Guide.YLabel("y-label"),
Guide.Title("Title of the plot"))
```

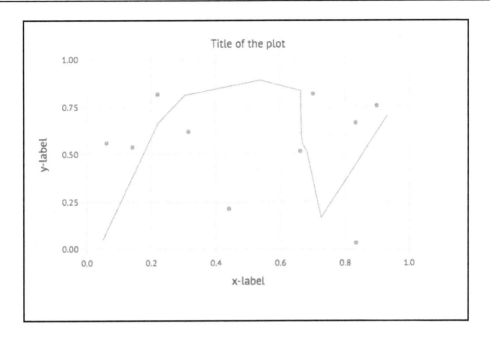

How it works...

In this section, we will look at plotting functions. Often, the data needs to be fitted and transformed into common functions, so having methods to plot them is very important. In the first example, the `sin` function has been plotted, which is a continuous function unlike the discrete plot, which we plotted for discrete data in the previous recipe, in the form of scatter plots and histograms.

Much as with histograms and scatter plots, we sometimes need to plot multiple functions in the same plot sometimes. This can be achieved by including the function information inside an array in the plot call. So, in the preceding second example, both the `sin` and `cos` functions have been plotted in the same plot. The start and end limits can be specified for the functions in the same plot call. The aesthetics for the multiple function lines will be differentiated automatically through different colors for each.

In cases where we do not want both the functions to be plotted as a line plot, we can further make use of the layers functionality. This allows us to interpret the plot as multiple layers. In the preceding example, we have plotted both continuous and discrete data in the form of a scatter plot and a continuous line plot in separate layers, respectively.

To make the plots more easy to interpret, we can further customize the aesthetics by adding guides and legends, which can give more information about the plot, such as the axis names, to the user. This can be done simply by adding the `Guide` parameter in the plot, like we did previously. So, we added the title of the plot and the name of the x and y axes in the `Guide` parameter. We can also add in-plot guides, but that is beyond the scope of the current chapter.

Exploratory data analytics through plots

Exploratory data analytics is one of the most important processes in a data science workflow. It is simply a thorough exploration of the data to find any possible patterns that can be identified through basic statistics and the shape of the data. It is mostly done with the help of plots, as visual information is much easier to comprehend than complex statistical terms. So, in this recipe, we will go through some exploratory analytics methods with the help of plots.

Getting ready

The `Gadfly` library, which we used for our recipes, also contains most of the plots that are frequently used for exploratory data analytics. We will use the same library for this purpose too. So, to install the library, you can follow the installation steps mentioned in the previous recipes.

We will also use datasets from the `RDatasets` package, which contains datasets that are in the data repository of the R programming language. So, to install the `RDatasets` package and invoke it, we follow the same steps we used previously:

```
Pkg.add("RDatasets")
using RDatasets
```

How to do it...

1. The first plot we will look at is the Box plot, which is used for identifying the spread of the distributions and also for analyzing outliers, which are basically anomalous data points in the data. We will use a dataset from the `RDatasets` package. For plotting a Box plot, we need to add just the `Geom.boxplot` inside the `plot()` function. This can be done as follows:

```
plot(dataset("lattice", "singer"), x = "VoicePart", y = "Height",
Geom.boxplot)
```

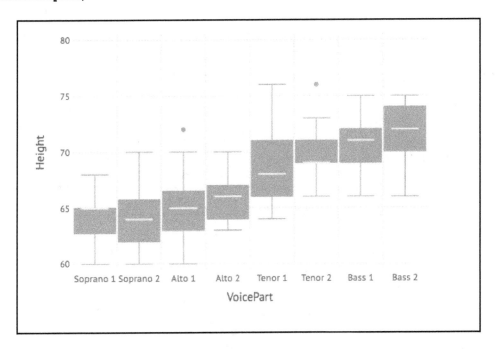

2. The next plot we will be looking at is the density plot (sometimes called the **frequency plot**). As the name suggests, it is a plot of the frequency of observations in a dataset. This plot is generally used to see how a particular column or dimension of a dataset is distributed. The syntax is the same as the one used previously, except now the aesthetic will be `Geom.density`. This can be done as follows:

```
plot(dataset("ggplot2", "diamonds"), x = "Price", Geom.density)
```

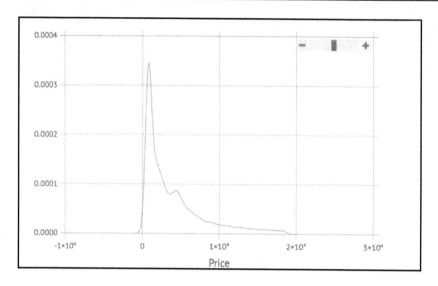

3. The next plot that we will explore is the bar plot, which is very popularly used for visualizing distributions, including the mode and the spread amongst most other properties. The plot can be created just like the other plot used previously, that is, by tweaking the aesthetics parameter to `Geom.bar`. This can be done as follows:

```
plot(dataset("HistData", "ChestSizes"), x = "Chest", y = "Count", Geom.bar)
```

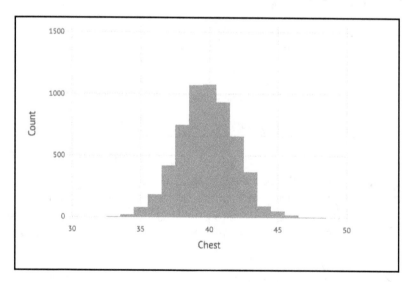

4. The final plot which we will look at in this recipe will be the histogram. This looks similar to the preceding bar plot, but is much more customizable, and it is used for a different purpose. The aesthetic parameter for this plot is `Geom.histogram`. It can be used as follows:

```
plot(dataset("ggplot2", "diamonds"), x = "Price", Geom.histogram)
```

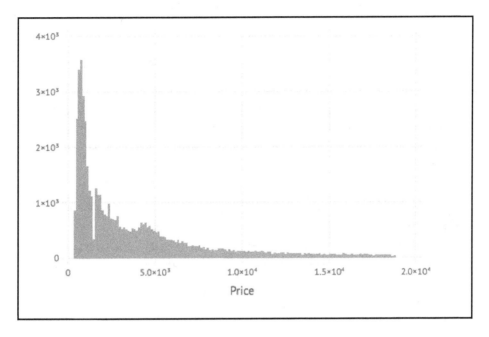

How it works...

If you observe the box plot properly, the lower and top whiskers in that plot are the minimum and maximum values (height) of that particular class. The white lines in the middle are the mean value of the class. The lower and upper parts of the boxes are the 25% and the 75% sections of the distribution. So, boxplots even help in the outlier detection and elimination process too.

The density plot shows the distribution or, as the name suggests, the density of a particular variable or a feature of a dataset. The preceding plot in the example is the frequency plot of the price column in the diamonds dataframe. It shows how the prices of the diamonds are distributed; from the plot, we can infer that they are a left-skewed distribution.

The bar plot is for plotting categorical variables against their count or some other metric. In the preceding example, we plotted the chest sizes of a population. So, from the bar plot, we can infer that chest sizes are normally distributed, which means that the three statistics of the data (the mean, median, and mode) all lie almost at the center of the distribution.

A histogram is another plot that shows the distribution of the data, and it is very closely related to density plots. Each bar is a bin and signifies the count of each range. The bin size and counts can be varied, which will be covered in later recipes in this chapter.

Line plots

Line plots, as we have already seen in the preceding examples, are very effective when it comes to exploratory data analytics. They can be used both to understand correlations and look at data trends. So, by further making use of aesthetics, we can make them more interesting and informative.

Getting ready

We will use the `Gadfly` library, which we have used in the preceding recipes. So, to install the library, you can follow the installation steps mentioned in the previous recipes.

How to do it...

1. Let's start with a basic line plot, which plots their incidences of melanoma in the respective years. So, this plot can be seen as a typical time series plot, where the x axis is a time variable and the y axis is the variable that is parameterized by time. So, to plot this, we simply need to include the dataset in the `plot()` function and include the `Geom.line` aesthetic, as follows:

```
plot(dataset("Lattice", "melanoma"), x = "Year", y = "Incidence",
Geom.line)
```

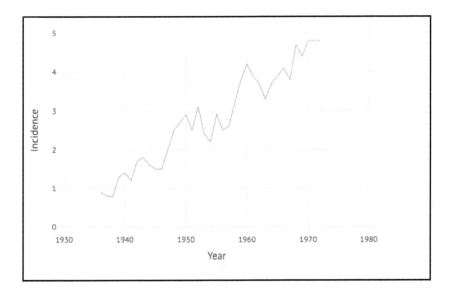

2. We can also have multiple line plots in the same plot. This will help us compare qtrends and the correlations across multiple columns or even across timelines. We can also use aesthetics to add separate colors to line plots. This can be done as follows:

```
plot(dataset("Zelig", "approval"), x = "Month", y = "Approve", color =
"Year", Geom.line)
```

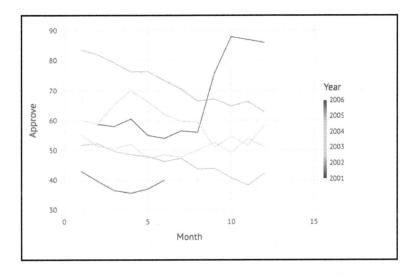

How it works...

A broken line plot is appropriate for visualizing time series data. The x axis represents the time dimension and the y axis represents the time-dependent feature. The plot in the first example shows a time series of the incidences of melanoma. We can see that the overall trend is increasing, but there are quite a lot of ups and downs, especially in the years 1958 and 1962.

We can also have multiple time series line plots in a single visualization by color-encoding each and putting up a legend for ease of interpretation. In the second plot, we compared the leave approval count of each month for 6 years. This helps us compare how the trends are changing with every year and also helps us identify possible patterns and cycles in the time series.

Scatter plots

Scatter plots are the most basic plots in exploratory analytics. They help the analyst get a rough idea of the data distribution and the relationship between the corresponding columns, which in turn helps identify some prominent patterns in the data.

Getting ready

We will use the `Gadfly` library, which we used in the preceding recipes. So, to install the library, you can follow the installation steps mentioned in the previous recipes.

How to do it...

1. Let's start off with plotting a simple scatter plot of iris features: the length and the width. This will help us identify the relationship between the two features of the flower. This can be done using a line plot similar to the one in the preceding recipe, but including the aesthetic `Geom.point` instead of `Geom.line` in the `plot()` function. This can be done as follows:

```
plot(dataset("datasets", "iris"), x = "SepalLength", y = "SepalWidth",
Geom.point)
```

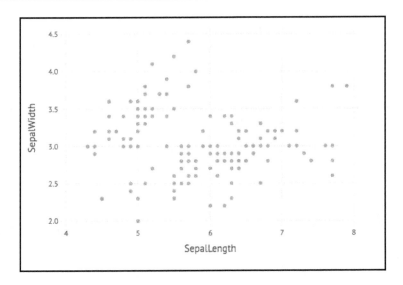

2. Next, we will try to put in some aesthetics on the plot to make it more informative. Coloring the plot with respect to flower species will give us good idea of the clusters and also the features of the individual species. This can be done as follows:

```
plot(dataset("datasets", "iris"), x = "SepalLength", y = "SepalWidth",
color = "Species", Geom.point)
```

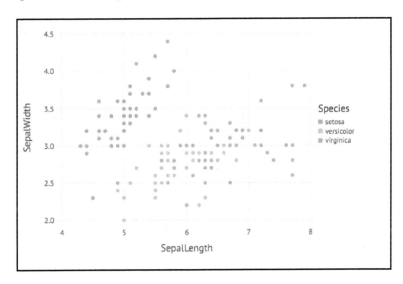

3. We can also use the color as a gradient indicator of any feature. For example, in this plot, we will use color as a gradient indicator of the petal length. Tweaking the color parameter to include the name of the feature will do. This can be done as follows:

```
plot(dataset("datasets", "iris"), x = "SepalLength", y = "SepalWidth",
color = "PetalLength", Geom.point)
```

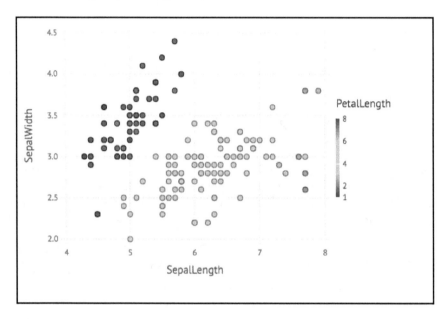

How it works...

Scatter plots are one of the most used visualization modes in the exploratory analytics part of an experiment. They are easy to interpret and are readily available in any visualization library. In the first of the earlier plots, we made a very basic scatter plot showing the length versus the width of the sepals of the iris flower.

The plots can be further customized to have separate colors for each type of iris. So, in the second plot, we can see the groups of each iris species encoded in a different color. This helps us see how each species compares with others when it comes to sepal length and the width. We can infer that the Sentosa species has a wider and a shorter sepal.

Plots can be further customized to add a gradient for a particular column so that the distribution can be easily identified. In the third plot, the gradient is applied across the length of the petal. The gradient's maximum and minimum values can also be customized, which will be covered in the final recipe of this chapter.

Histograms

Histograms are one of the best ways for visualizing and finding out the three main statistics of a dataset: the mean, median, and mode. Histograms also help analysts get a very clear understanding of the distribution of data. The ability to plot categorical data as well as numerical data is what makes the histogram unique.

Getting ready

We will use the Gadfly library, which we used for understanding and plotting data in the preceding recipes. So, to install the library, you can follow the installation steps mentioned in the previous recipes.

How to do it...

1. A basic histogram is a simple set of stacked bars, which shows the distribution of a particular feature in a dataset. This can be plotted using the plot() function, with the Geom.histogram attribute as the aesthetic parameter. We will use the diamonds dataset for the purpose. This can be done as follows:

```
plot(dataset("ggplot2", "diamonds"), x = "Price", Geom.histogram)
```

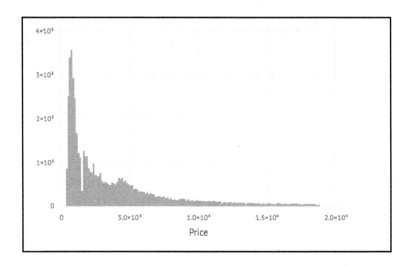

2. As with earlier plots, color aesthetics can be used to differentiate the categories in the data. There is a `color` parameter in the `plot` definition, which takes in a categorical feature of the dataset for color encoding. This can be done as follows:

```
plot(dataset("ggplot2", "diamonds"), x = "Price", color = "Cut",
Geom.histogram)
```

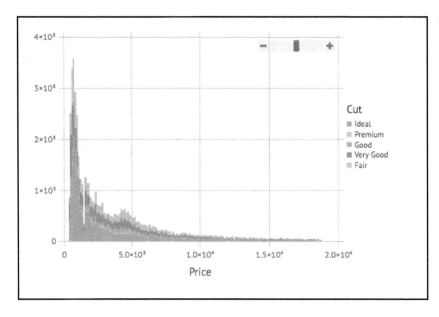

3. One very important part of a histogram is the binning feature, which is the count

of data points that can fall into a bin in the *x* axis. This will determine the ease of interpretation of the data. We will try to reduce the number of bins in this example using the `bincount` parameter in the `Geom.histogram` aesthetic. This can be done as follows:

```
plot(dataset("ggplot2", "diamonds"), x = "Price", color = "Cut"
Geom.histogram(bincount = 45))
```

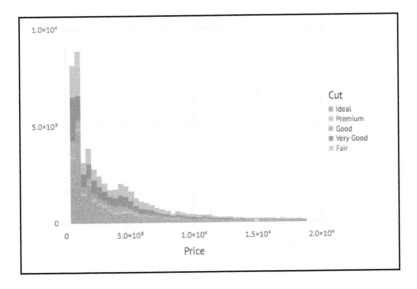

How it works...

Histograms are very commonly used for studying distributions of data features. The three basic statistics of data are a worthy study study in a histogram plot. The *y* axis shows the number of elements that fall into each bin of the *x* axis.

Also, the stacks can be color-coded with respect to a categorical variable in a particular feature. In the second example, it is the type of the diamond that is color-coded into the histogram; we can clearly see that premium and very good diamonds are priced higher than the other categories available.

The bin size can be customized, which means that the ranges of the bins can be further increased and decreased according to ease of interpretation. In the preceding example, the bin sizes have been increased, which means that the range of each bin is increased. And that helps us look at the histogram and better perceive the data distribution.

Aesthetic customizations

As we have already gone through how to plot the most important visualizations and their customizations in the Gadfly library, we will also see how to customize them even further. The Gadfly library allows the analyst to almost completely tweak and customize their visualizations so that they can be better fitted to the dataset properties are very flexible for our purposes.

Getting ready

We will use the Gadfly library, which we used in the preceding recipes. So, to install the library, you can follow the installation steps mentioned in the previous recipes.

How to do it…

1. The limits of the axes can be customized or transformed to the logarithmic scale with the Scale.x_log parameter in the plot() function. This would help in visualizing exponentially increasing data or data in different scales. We will scale the x axis in this example. This can be done as follows:

```
plot(x = rand(10), y = rand(10), Scale.x_log)
```

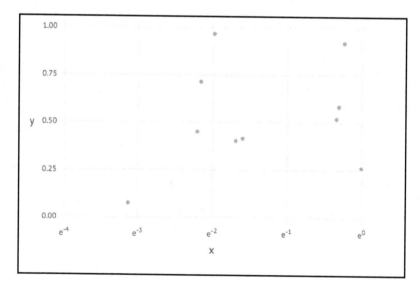

2. The minimum and maximum values in the plot or in a particular axis can also be set to custom values when we want to slice the dataset or scale a feature for better visualizing. In this example, we will scale the *x* axis. This can be done as follows:

```
plot(x = rand(10), y = rand(10), Scale.x_continuous(minvalue = -20,
maxvalue = 20))
```

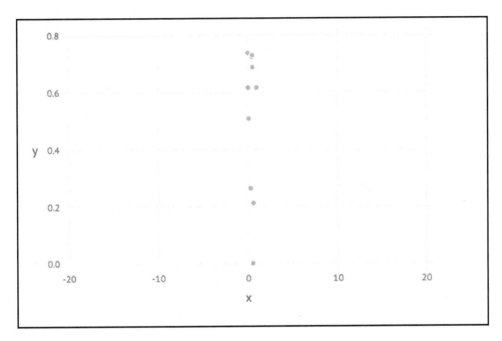

3. Discrete data can also be scaled in a similar way. The `Scale.x_discrete` parameter can be used for the purpose of scaling discrete data in the *x* axis, as follows:

```
plot(x = rand(1:5, 10), y = rand(10), Scale.x_discrete)
```

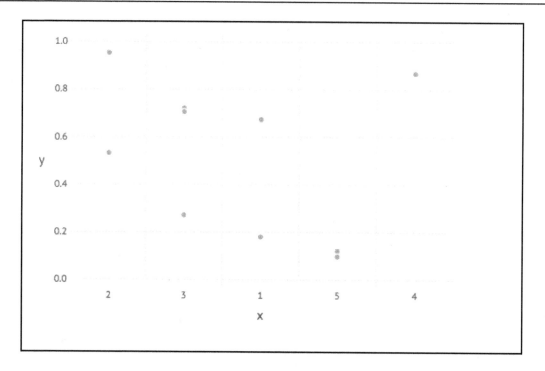

How it works...

Gadfly gives the analyst a lot of flexibility when it comes to customizing plots. In the first plot, we used the transformation feature to scale the data logarithmically for better comparison. Sometimes, data might be exponentially distributed, and it might not be possible to visualize it in a single plot. So, in those circumstances, scaling it to a logarithmic scale will help the analyst fit it in a single plot.

As pointed out in several examples in earlier recipes, the minimum and maximum values for the axes can also be customized. This helps us slice and dice the plots according to our requirements, just like we have been able to slice and dice dataframes depending on specific conditions.

Discrete data can be easily plotted and scaled accordingly, too. The third plot showed a plot of discrete data, scaled through a Gadfly customization. All the preceding plots have been customized for the *x* axis. The *y* axis can also be customized similarly by replacing the *x* axis with the *y* axis in the plot () function or by adding an extra *y* axis customization along with that of the *x* axis.

6
Parallel Computing

In this chapter, we will cover the following recipes:

- Basic concepts of parallel computing
- Data movement
- Parallel map and loop operations
- Channels

Introduction

In this chapter, you will learn about performing parallel computing and using it to handle big data. So, some concepts such as data movements, sharded arrays, and the Map-Reduce framework are important to know in order to handle large amounts of data by computing on it using parallelized CPUs. So, all the concepts discussed in this chapter will help you build good parallel computing and multiprocessing basics, including efficient data handling and code optimization.

Basic concepts of parallel computing

Parallel computing is a way of dealing with data in a parallel way. This can be done by connecting multiple computers as a cluster and using their CPUs to carry out the computations.

This style of computation is used when handling large amounts of data and also while running complex algorithms over significantly large data. The computations are executed faster due to the availability of multiple CPUs running them in parallel as well as the direct availability of RAM to each of them.

Getting ready

Julia has in-built support for parallel computing and multiprocessing. So, these computations rarely require any external libraries for the task.

How to do it...

1. Julia can be started on your local computer using multiple cores of your CPU. So, we will now have multiple workers for the process. This is how you can fire up Julia in multi-processing mode in your terminal. This creates two worker process in the machine, which means it uses two CPU cores for the purpose:

```
julia -p 2
```

The output looks something like this. It might vary for different operating systems and different machines:

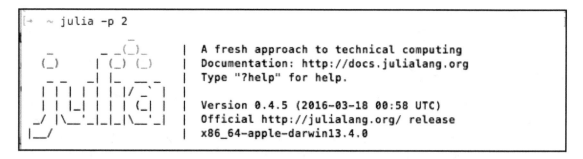

2. Now, we will look at the `remotecall()` function. It takes in multiple arguments, the first one being the process which we want to assign the task to. The next argument will be the function we want to execute. The subsequent arguments will be the parameters or the arguments of the function we want to execute. In this example, we will create a 2 x 2 random matrix and assign it to the process number 2. This can be done as follows:

```
task = remotecall(2, rand, 2, 2)
```

The preceding command gives the following output:

```
[julia> task = remotecall(2, rand, 2, 2)
RemoteRef{Channel{Any}}(2,1,3)
```

3. Now that the `remotecall()` function for remote referencing has been executed, we will fetch the results of the function through the `fetch()` function. This can be done as follows:

```
fetch(task)
```

The preceding command gives the following output:

```
[julia> fetch(task)
2x2 Array{Float64,2}:
 0.0282701  0.37992
 0.116476   0.833553
```

4. Now, to perform some mathematical operations on the generated matrix, we can use the `@spawnat` macro, which takes in the mathematical operation and the `fetch()` function. The `@spawnat` macro actually wraps the expression `5 .+ fetch(task)` into an anonymous function and runs it on the second machine This can be done as follows:

```
task2 = @spawnat 5 .+ fetch(task)
```

```
[julia> task2 = @spawnat 2 5 .+ fetch(task)
RemoteRef{Channel{Any}}(2,1,5)

[julia> fetch(task2)
2x2 Array{Float64,2}:
 5.02827  5.37992
 5.11648  5.83355
```

5. There is also a function that eliminates the need for using two different functions: `remotecall()` and `fetch()`. The `remotecall_fetch()` function takes in multiple arguments. The first one is the process that the task is being assigned. The next argument is the function you want to be executed. The subsequent arguments will be the arguments or the parameters of the function that you want to execute. Now, we will use the remote `call_fetch()` function to fetch an element of the task matrix for a particular index. This can be done as follows:

```
remotecall_fetch(2, getindex, task2, 1, 1)
```

```
[julia> remotecall_fetch(2, getindex, task2, 1, 1)
5.028270064003363
```

How it works…

Julia can be started in multiprocessing mode by specifying the number of processes needed while starting up the REPL. In this example, we started Julia as a two-process mode. The maximum number of processes depends on the number of cores available in the CPU.

The `remotecall()` function helps in selecting a particular process from the running processes in order to run a function or, in fact, any computation for us.

The `fetch()` function is used to fetch the results of the `remotecall()` function from a common data resource (or the process) for all the running processes. The details of the data source will be covered in later sections.

The results of the `fetch()` function can also be used for further computations, which can be carried out with the `@spawnat` macro along with the results of `fetch()`. This will assign a process for the computation.

The `remotecall_fetch()` function further eliminates the need for the fetch function in the case of a direct execution. This has both the `remotecall()` and `fetch()` operations built into it. So, it acts as a combination of both the second and third points in this section.

Data movement

In parallel computing, data movements are quite common and should be minimized due to the time and the network overhead as a result of the movements. In this recipe, we will see how that can be optimized to avoid latency as much as we can.

Getting ready

To get ready for this recipe, you need to have the Julia REPL started in multiprocessing mode. This is explained in the *Getting ready* section in the preceding recipe.

How to do it...

1. Firstly, we will see how to do a matrix computation using the @spawn macro, which helps in data movement. So, we construct a matrix of shape 200 x 200 and then try to square it using the @spawn macro. This can be done as follows:

```
mat = rand(200, 200)
exec_mat = @spawn mat^2
fetch(exec_mat)
```

The preceding command gives the following output:

```
julia> mat = rand(200, 200)
200x200 Array{Float64,2}:
 0.649012   0.871952   0.108619   …  0.851265   0.0502902  0.218886
 0.920093   0.658257   0.438752      0.142323   0.726779   0.172245
 0.0782007  0.270488   0.0175621     0.837172   0.29039    0.278585
 0.0168289  0.182208   0.214846      0.337172   0.679423   0.642266
 0.229643   0.954157   0.359372      0.953347   0.760825   0.19385
 0.990622   0.885887   0.343064   …  0.837153   0.219632   0.23556
 0.800763   0.552112   0.452425      0.265355   0.789556   0.46955
 0.0972477  0.07953    0.945542      0.57119    0.604993   0.71299
 0.411358   0.61751    0.604449      0.255039   0.666815   0.282209
 0.246472   0.90213    0.0890926     0.905304   0.228333   0.492838
 ⋮                                ⋱
 0.247632   0.368542   0.567844      0.0147391  0.180392   0.228339
 0.839398   0.0990057  0.663658      0.614794   0.649242   0.709225
 0.647001   0.132353   0.200982      0.295815   0.0221692  0.00974389
 0.467503   0.546979   0.157562      0.00776156 0.742553   0.0624323
 0.700403   0.363153   0.642075   …  0.700312   0.96451    0.27512
 0.637208   0.350748   0.389078      0.490583   0.499495   0.466656
 0.702178   0.677843   0.122308      0.830219   0.926329   0.00571204
 0.818075   0.396422   0.40664       0.575228   0.663198   0.375351
 0.944022   0.340856   0.396091      0.544566   0.968218   0.961706

julia> exec_mat = @spawn mat^2
RemoteRef{Channel{Any}}(2,1,9)

julia> fetch(exec_mat)
200x200 Array{Float64,2}:
 50.3471  48.5066  50.2064  49.541   …  46.3003  52.5357  51.326   46.6558
 52.4674  49.2354  50.8089  48.0996     46.4235  47.6066  53.1069  46.6121
 52.9261  50.9873  50.9772  49.4982     46.9337  52.675   54.0344  49.8417
 48.7711  46.9278  49.2939  44.2167     45.2764  48.8136  50.7424  43.1134
 48.2666  49.3658  49.5425  47.7228     44.0711  48.3831  48.7698  44.389
 53.3778  54.7359  51.7429  51.1009  …  49.6738  53.1695  53.4594  47.4184
 49.4719  48.3891  52.1893  46.7813     47.3374  50.8025  49.2288  46.1703
 49.4079  49.5932  52.062   47.7079     47.4241  52.2929  50.7447  47.3584
 48.1788  46.6472  48.9192  45.1642     43.4388  50.2002  50.1279  45.9442
 50.4437  46.9865  48.3177  48.331      46.2023  50.7359  52.0813  45.3443
 ⋮                                   ⋱
 46.9014  47.481   47.0781  42.099      43.3535  48.7576  48.9715  43.4123
 52.8539  50.2219  52.4422  49.8037     47.0852  51.162   53.6264  49.2556
 46.7897  45.229   48.2966  45.3931     45.7173  46.5399  47.6638  44.6588
 49.4582  49.9821  48.925   48.6939     47.6608  49.1771  51.3251  46.2242
 50.8366  52.6709  50.7749  49.7035  …  47.3265  51.2981  51.6807  47.1084
 52.43    49.3027  50.3005  48.7467     49.5461  49.805   50.9799  47.8435
 48.6858  47.4298  49.9422  45.8484     42.702   48.1379  48.8334  43.0755
 50.4815  51.1354  50.3583  49.6866     50.0998  52.2442  52.836   49.8903
 51.9598  52.6884  50.4821  49.4032     51.3082  53.0197  52.235   49.9443
```

2. Now, we will look at an another way to achieve the same result. This time, we will use the @spawn macro directly instead of the initialization step. We will discuss the advantages and drawbacks of each method in the *How it works...* section. So, this can be done as follows:

```
mat = @spawn rand(200, 200)^2
fetch(mat)
```

The preceding command gives the following output:

```
julia> mat = @spawn rand(200, 200)^2
RemoteRef{Channel{Any}}(3,1,11)

julia> fetch(mat)
200x200 Array{Float64,2}:
54.5697  53.8363  53.6856  51.0336  …  51.0487  51.4376  57.5861  52.3218
47.2254  44.2958  45.6051  44.2316     42.866   44.7235  49.2066  45.3419
48.7921  47.206   49.3328  45.8813     47.133   48.8372  50.9881  49.4069
53.3333  52.2756  51.0586  48.1974     50.611   49.3214  54.932   49.3784
45.9031  46.1887  47.2085  44.2307     43.1819  46.0402  48.7807  43.999
52.047   53.3896  52.5301  49.7597  …  49.9737  53.3288  57.1498  52.375
45.7131  45.6933  48.2079  41.9669     46.9972  45.8021  51.2106  47.2518
47.5226  47.719   47.8226  45.6203     45.2402  47.358   49.5541  44.3646
50.1499  47.7953  48.169   47.4005     47.0003  48.2151  52.6357  46.6296
  ⋮                                ⋱
46.7402  45.4074  47.9322  43.4595     45.9452  48.9702  51.1098  46.3291
48.8216  47.3026  45.3776  45.1829     44.561   45.7782  50.4167  45.4056
52.8919  51.3057  52.8045  48.2534     49.9482  50.8668  56.6768  52.1277
50.1029  49.5572  48.258   47.1801     47.1399  48.8574  48.8144  49.672
51.5512  51.3725  49.8671  48.8818  …  49.5948  51.8613  56.2478  51.0256
55.462   54.3591  51.9077  51.2633     51.0253  53.4258  57.4629  52.1847
49.5583  50.8754  48.9668  49.5569     45.6223  50.3501  52.0695  49.0455
57.2287  54.3223  55.7038  52.2581     52.9821  53.8025  59.5563  53.8772
47.5443  48.9313  49.4089  45.0021     49.5035  49.2289  53.7818  47.8232
```

How it works...

In this example, we try to construct a 200 X 200 matrix and then used the @spawn macro to spawn a process in the CPU to execute this for us. The @spawn macro spawns one of the two processes running, and it uses one of them for the computation.

In the second example, you learned how to use the @spawn macro directly without an extra initialization part. The fetch() function helps us fetch the results from a common data resource for the processes. We will cover this in more detail in the following recipes.

Parallel maps and loop operations

In this recipe, you will learn a bit about the famous Map-Reduce framework and why it is one of the most important ideas in the domains of big data and parallel computing. You will learn how to parallelize loops and use reducing functions on them through several CPUs and machines and you will further explore the concept of parallel computing, which you learned about in the previous recipes.

Getting ready

Just like the previous sections, Julia just needs to be running in multiprocessing mode to work through the following examples. This can be done through the instructions given in the first section.

How to do it...

1. Firstly, we will write a function that takes and adds n random bits. The writing of this function has nothing to do with multiprocessing. So, it has simple Julia functions and loops. This function can be written as follows:

```
[julia> function count_heads(n)
           c::Int = 0
           for i = 1+n
               c += rand(Bool)
           end
           c
       end
count_heads (generic function with 1 method)
```

2. Now, we will use the @spawn macro, which we learned about previously, to run the count_heads() function as separate processes. The count_heads() function needs to be in the same directory for this to work. This can be done as follows:

```
require("count_heads")
a = @spawn count_heads(100)
b = @spawn count_heads(100)
fetch(a) + fetch(b)
```

3. However, we can use the concept of multi-processing and parallelize the loop directly; we can also sum it. The parallelizing part is called **mapping**, and the addition of the parallelized bits is called**reduction**. Thus, the process constitutes the famous Map-Reduce framework. This is made possible using the `@parallel` macro, as follows:

```
nheads = @parallel (+) for i = 1:200
    Int (rand(Bool))
end
```

```
julia> nheads = @parallel (+) for i = 1:200
[          Int(rand(Bool))
[      end
92
```

How it works...

The first function is a simple Julia function that adds random bits with every loop iteration. It was created just for the demonstration of Map-Reduce operations.

In the second point, we spawn two separate processes for executing the function; we then fetch the results of both of them and add them up.

However, that is not really a neat way to carry out parallel computation of functions and loops. Instead, the `@parallel` macro provides a better way to do it; it allows the user to parallelize the loop and then reduce the computations through an operator, which together constitutes the Map-Reduce operation.

Channels

Channels are like background plumbing for parallel computing in Julia. They are the reservoirs from which the individual processes access their data.

Getting ready

The requirements are similar to the previous sections. This is mostly a theoretical section, so you just need to run your experiments on your own. For that, you need to run your Julia REPL in a multiprocessing mode.

How to do it...

Channels are shared queues with a fixed length. They are common data reservoirs for the processes which are running.

The channels are like common data resources, which multiple readers or workers can access. They can access the data through the `fetch()` function, which we already discussed in the previous sections.

The workers can also write to the channel through the `put!()` function. This means that the workers can add more data to the resource, which can be accessed by all the workers running a particular computation.

Closing a channel after use is a good practice to avoid data corruption and unnecessary memory usage. It can be done using the `close()` function.

Index

www.ingramcontent.com/pod-product-compliance
Lightning Source LLC
LaVergne TN
LVHW081344050326
832903LV00024B/1296